Library of
Davidson College

Higher Oil Prices:
Worldwide Financial Implications

A Policy Statement
   by the British-North American Committee

A Research Report
   by Sperry Lea

BRITISH-NORTH AMERICAN COMMITTEE

Sponsored by
British-North American Research Association
National Planning Association (U.S.A.)
C.D. Howe Research Institute (Canada)

The British-North American Committee

At the Committee's first meeting in New York City, December 1969, the following statement of its aims was authorized:

> The British-North American Committee has been established to study and comment upon the developing relationships between Britain, the United States and Canada. It seeks to promote clearer understanding of the economic opportunities and problems facing the three countries, to explore areas of cooperation and of possible friction, and to discover constructive responses. It believes that sound relations between these three countries in the context of an increasingly interrelated world are essential to future prosperity and seeks to promote better understanding through the collection of facts and their widespread dissemination.

In serving these aims, the Committee is sponsoring a series of objective studies undertaken by qualified experts in the three countries and published with the Committee's approval. On the basis of these factual studies and of discussions at its meetings, the Committee may also issue policy statements signed by its members.

The Committee's membership—listed on pp. 32-35—includes business, banking, labor, agricultural, and professional leaders from Britain, the United States and Canada. The Committee is sponsored by three nonprofit research organizations—the British-North American Research Association in London, the National Planning Association in Washington, and the C.D. Howe Research Institute in Montreal, described on pp. 36-37.

The British-North American Committee is a unique organization both in terms of its broadly diversified membership and in terms of its blending of factual studies and policy conclusions on British-North American relations. It meets twice a year, once in Great Britain and once in North America. Its work is jointly financed by funds contributed from private sources in Canada, Great Britain and the United States.

Offices on behalf of the Committee are maintained at 1606 New Hampshire Avenue N.W., Washington, D.C. 20009, and at 6/14 Dean Farrar Street, London, SW1H ODX. George Goyder serves as British Secretary of the Committee and John Miller (Assistant Chairman and President of the National Planning Association) serves as North American Secretary. Simon Webley in London and Sperry Lea in Washington are the Co-Directors of Research.

DAVID BARRAN   NICHOLAS J. CAMPBELL, Jr.

Co-Chairmen of the Committee

Higher Oil Prices:
Worldwide Financial Implications

A Policy Statement
by the British-North American Committee

A Research Report
by Sperry Lea
National Planning Association

BRITISH-NORTH AMERICAN COMMITTEE

Sponsored by
British-North American Research Association
National Planning Association (U.S.A.)
C.D. Howe Research Institute (Canada)

© British-North American Committee 1975
Short quotations with appropriate credit permissible

SBN 902594-27-3
Library of Congress Catalog Card Number 75-29675

Published by the British-North American Committee
Printed and bound in the U.S.A.

October 1975

> This publication on the worldwide financial implications of higher oil prices comprises two elements: a Policy Statement in which the Committee speaks out in its own voice, and a Research Report on the same subject. The Statement was adopted by the Committee during a regular semi-annual meeting, at Bournemouth, England, in the summer of 1975. Signatures and footnotes of members endorsing it are appended.
>
> The Research Report was prepared by Sperry Lea, the Committee's North American Research Director, who is responsible for its contents. An Appendix, compiled separately in the BNAC's London office by Susan Hart, follows Sperry Lea's paper.

Contents

The British-North American Committee .... inside front cover

**A Policy Statement by the
British-North American Committee** v

Members of the Committee Signing the Statement viii

**Worldwide Financial Implications of Higher Oil Prices
by Sperry Lea** 1

Introduction 1

   I. What Are We Talking About? 3
      Dramatis Personae 3
      The Basic Division of OPEC Oil Revenues 4
      Measures of OPEC Surpluses 5
      Recycling 6

   II. Initial Perceptions and Reactions 9
      Improved Recycling 12
      Other Beneficial Policies and Developments 16

   III. Perceptions to 1980 18
      Contradictory Signals for OPEC Prices 18
      A New Phase in the Problems Facing
         Industrialized Countries 19
      The Situation Facing the LDCs 21

**Appendix: The Disposition of OPEC Oil Revenues
by Susan Hart** 25

Members of the British-North American Committee 32

Sponsoring Organizations 36

Publications of the British-
   North American Committee .... inside back cover

Contents

Exhibits .... center section

... for the Research Report

1. A Classification of the (Net) Oil-Importing Countries
2. The "Most Seriously Affected Countries" (MSAs)
3. Some Basic Data on OPEC Countries
4. Estimated OPEC Investible Surplus for 1975, with Summary Data for 1973 and 1974
5. OPEC Oil Prices, 1970-74
6. Primary Recycling: Estimated Deployment of OPEC Surpluses, 1974 and 1975
7. OPEC Aid: Commitments and Disbursements in 1974
8. Modes of Recycling Surplus Petrodollars
9. The Spectrum of Recycling to Governments
10. The Operations of the IMF Oil Facilities
11. OPEC Current Account Estimates
12. Publicized Eurocurrency Credits: 1974 and First Half of 1975

... for the Appendix

Chart I. Main OECD Countries' Exports to OPEC, 1970-74

Table I. Value of Exports to OPEC Countries from Selected OECD Countries, 1973 & 1974

Table II. Some Middle Eastern OPEC Country Defence Expenditure Contracts Reported in 1974-75

A Policy Statement by the British-North American Committee on Higher Oil Prices: Worldwide Financial Implications

The sudden fivefold increase in the Organization of Petroleum Exporting Countries' (OPEC) oil prices in the winter of 1973-74, coupled with the temporary selective Arab oil embargo, confronted virtually all nations with their single greatest task of economic adjustment since the aftermath of World War II. Besides exerting a substantial recessionary drag and adding significantly to the inflationary impetus already existing in the consuming countries, the higher oil prices were seen to pose two immediate financial challenges:
●Could sufficient funds for borrowing be made available to those countries unable to finance from other sources their balance-of-payments deficits arising from higher oil costs?
●Could the private international financial institutions fulfill their role as the principal intermediaries, absorbing unprecedented inflows of surplus OPEC oil revenues and relending them without exposure to excessive risk?

At the same time, the oil price increase generated some medium-term problems with which the consuming countries must grapple.

In 1974, the fear was that, if the immediate challenges were not met, countries with oil-induced financing problems might seek rapid readjustment through restricting imports and slowing down economic activity, or by concluding special bilateral deals with individual OPEC members, or both—hasty unilateral steps of dubious benefit to those adopting them and of certain harm to many other oil-importing countries. Furthermore, the private international financial institutions might cease to absorb funds offered, thus giving an incentive to producing nations to reduce their financial surpluses by reducing the quantity of oil produced.

During the past 18 months, however, only a few such undesirable responses have materialized, largely owing to the effective operation of market forces and to the success of cooperative actions. The latter include new mechanisms—created by agreements among consuming countries, among OPEC countries, or involving both groups—for recycling OPEC surpluses to needy countries and underpinning the banking system.

Meanwhile, the actual oil-induced deficits of the consuming countries are lower than expected owing to reduced demand resulting from two mild winters, the worldwide recession and higher oil prices. Furthermore, higher than expected OPEC imports are considered, by some, to indicate a trend which will continue. This factor, plus reappraisals of future oil prices and of the effectiveness of conservation measures, raises the possibility that the accumulation by OPEC of surplus petrodollars may well be considerably lower by the end of the 1970s than was forecast in the summer of 1974.

Lessons

The British-North American Committee considers that a careful examination of the events and the responses to them during the past 18 months can provide a clear guide for dealing with future problems arising from the higher oil prices. The Committee draws two particular lessons from recent history.

First, *market forces* have by themselves brought about considerable beneficial readjustment.

During a period when governments have been criticized frequently for acting slowly and indecisively to solve their recycling problems, market forces have been operating in a number of ways to bring about natural adjustments in desirable directions. For instance, oversupply of liquid capital placed by OPEC countries in the world banking system contributed to the decline of short-term interest rates, thus encouraging OPEC investment into longer-term or fixed assets. Market forces have also helped to stabilize real oil prices and are exerting some downward pressure on them. Furthermore, market forces have stimulated a serious search for alternative energy sources and for more efficient uses of present energy supplies.

In the light of these developments, *the Committee urges governments to recognize and respect the potential of market forces to accomplish some of the most difficult and important tasks of adjustment.* We suggest that governments, in response to higher oil prices, should focus on conservation measures, encouraging the development of new energy supplies and collaborative efforts with other countries.

Second, when world economic stability is threatened, *international cooperation* works.

Insofar as governments have acted effectively, they have contributed to the successful adjustment of the financing problem through joint actions by various country groupings. For instance, OPEC countries and other International Monetary Fund (IMF) members have participated in setting up its special oil facilities; OPEC countries are taking increasing responsibility for helping developing countries; central banks have agreed to buttress banks operating in the Eurodollar market, the key instruments of recycling; European Community (EC) members have established their own loan fund facility; and the Organization for Economic Cooperation and Development (OECD) nations (minus France) have put into action the International Energy Agency and all of them are presently setting up their $25 billion Financial Support Fund.

The Future

Because the habit of international consultation and the discipline of having to adjust to the point of view of others has been enhanced by responding to the "oil crisis," *the Committee urges individual governments not to pursue unilateral solutions to some medium-term problems that are still to be faced.*

Three major examples of those which the Committee sees arising are:

(1) Surplus revenues arising from higher oil prices will continue to accumulate in a few oil-producing countries. These surpluses, which will be very large in terms of our previous experience, will require smooth recycling to the deficit countries. The Committee considers, however, that the relative success so far achieved by the international monetary authorities in containing the problems of recycling could well be transitory and that intensified international cooperation, especially between oil-producing and oil-consuming nations, will certainly be needed to preserve stability in the world financial and economic system.

(2) OPEC countries are presently investing in financial assets of one form or other close to $50 billion annually in the oil-consuming countries. The accumulation of these "surplus petrodollars," together with unremitted interest and dividends, constitutes massive indebtedness of the importing countries that sooner or later they must repay by exports of goods and services to OPEC countries at far above present levels. The potential size of such "transfers of real resources" is enormous—several hundred billion dollars by the end of the decade. If this shift in the consuming countries from incurring immense debts to providing goods occurs extensively just when their production facilities were again becoming fully occupied in meeting normal orders, then "demand-pull" inflation would be generated which might reinforce a residual "cost-push" inflation. To accommodate increasing demand without encouraging a further surge of inflation or rapid reduction in real incomes, the oil-consuming countries need to increase productive capacity.

The Committee therefore urges the developed consuming countries to find means to channel recycled petrodollars deposited with them into productive investment and not merely into financing existing or higher levels of consumption by underwriting deficits on current accounts.

(3) The position of the developing countries most seriously affected by the sharp rise in prices, not only for oil but also for food and fertilizer imports, has become critical. Although some special facilities have been set up to finance their growing deficits and a number of oil-producing countries have helped some of the most vulnerable developing countries, the problem is far from solved. Failure to recognize the importance of this problem, and to adopt a multilateral approach to it, will result in both growing economic deprivation and intensification of political instability in certain parts of the world. *The Committee strongly recommends that the plight of this particular group of countries be given high priority by both oil-producing and oil-consuming nations, and that discussions, perhaps with less public attention focused on them, be restarted as soon as possible. We urge the governments of the United Kingdom, the United States and Canada to take a lead in this matter.*

Footnote to the Statement

The "market forces" which helped reduce the financial strain of the surplus petrodollars included a substantial increase in sales of arms by the United States to OPEC countries. In the fiscal year ending June 30, 1974, arms sales abroad doubled, the bulk of the $8.5 billion total going to Middle East and Persian Gulf countries. Iran alone took $4 billion and Saudi Arabia $700 million.

That is not the kind of "market forces" to be encouraged, even though it greatly eases the financial problem of recycling oil money. Instead, as this Statement properly emphasizes, *productive* investment, in fertilizer plants, for example, should be pushed, along with aid to developing countries suffering from high oil and other import costs—**Lauren Soth**.

Members of the Committee Signing the Statement

Chairmen
SIR DAVID BARRAN
A Managing Director, Shell Transport
and Trading Company Ltd.

NICHOLAS J. CAMPBELL, JR.
Director and Senior Vice President,
Exxon Corporation

Vice Chairmen
IAN MacGREGOR
Chairman, AMAX Incorporated

RICHARD DOBSON
Chairman, British-American
Tobacco Co., Ltd.

Chairman, Executive Committee
W.O. TWAITS
Director and Vice President, Royal
Bank of Canada

Members
A.E. BALLOCH
Executive Vice President,
Bowater Incorporated

ROBERT BELGRAVE
Policy Planning Advisor,
British Petroleum Limited

RUSSELL BELL
Director of Research, Canadian Labour
Congress

I.H. STUART BLACK
Chairman, General Accident Fire and
Life Assurance Corporation Ltd.

HOWARD BLAUVELT
Chairman and Chief Executive Officer,
Continental Oil Company

HARRY BRIDGES
President and Chief Executive Officer,
Shell Oil Company

DR. CHARLES CARTER
Vice-Chancellor, University of Lancaster

SILAS S. CATHCART
Chairman and Chief Executive Officer,
Illinois Tool Works Inc.

HAROLD van B. CLEVELAND
Vice President, First National City
Bank

KIT COPE
Overseas Director, Confederation of
British Industry

WILLIAM DODGE
Ottawa, Ontario

ALASTAIR F. DOWN
Chairman and Chief Executive, Burmah
Oil Company

G. EASTWOOD
General Secretary, Association of
Patternmakers and Allied Craftsmen

LORD FEATHER
London

Committee Signers

JOSEPH B. FLAVIN
Executive Vice President, Xerox
Corporation

ROBERT M. FOWLER
President, C.D. Howe Research Institute

WILLIAM FRASER
Chairman, British Insulated
Callender's Cables Ltd.

DOUGLAS R. FULLER
Vice-Chairman, The Northern
Trust Company

GEORGE GOYDER
British Secretary, British-North
American Committee

EDWARD GRUBB
Chairman and Chief Officer,
The International Nickel Company
of Canada, Ltd.

HARRY G. JOHNSON
Professor of Economics, University
of Chicago

JOSEPH D. KEENAN
International Secretary, International
Brotherhood of Electrical Workers,
AFL-CIO

TOM KILLEFER
Vice President, Finance and General
Counsel, Chrysler Corporation

CURTIS M. KLAERNER
Executive Vice President and
Director, Mobil Oil Corporation

H.U.A. LAMBERT
Vice-Chairman, Barclays Bank Ltd.

HERBERT H. LANK
Director, Du Pont of Canada Ltd.

JOHN LAWRENCE
Chairman of the Board, Dresser
Industries, Inc.

FRANKLIN A. LINDSAY
Chairman of the Board, Itek
Corporation

JAMES LONGMORE
Director of Lloyds Bank, International
Limited

JAY LOVESTONE
International Affairs Consultant,
AFL-CIO

RAY W. MACDONALD
Chairman and Chief Executive
Officer, Burroughs Corporation

B.J. McGILL
Vice President and General Manager,
International, The Royal Bank of
Canada

DONALD E. MEADS
Chairman and Chief Executive Officer,
Certain-Teed Products Corporation

SIR PETER MENZIES
Chairman, The Electricity Council

DEREK F. MITCHELL
President, BP Canada Ltd.

JOSEPH P. MONGE
Vice Chairman, International Paper
Company

D.R. MONTGOMERY
Secretary-Treasurer, Canadian
Labour Congress

DR. MALCOLM MOOS
Santa Barbara, California

JOSEPH MORRIS
President, Canadian Labour Congress

CHARLES MUNRO
President, Canadian Federation of
Agriculture

KENNETH D. NADEN
President, National Council of Farmer
Cooperatives

JOSEPH NICKERSON
Chairman, The Nickerson Group of
Companies

WILLIAM S. OGDEN
Executive Vice President,
The Chase Manhattan Bank

WILLIAM R. PEARCE
Vice President, Cargill Incorporated

SIR RICHARD POWELL
Director, Hill Samuel Group Limited

Committee Signers

J.G. PRENTICE
Chairman of the Board, Canadian Forest Products Ltd.

BEN C. ROBERTS
Professor of Industrial Relations, London School of Economics

HAROLD B. ROSE
Group Economic Adviser,
Barclays Bank Limited,
London

WILLIAM SALOMON
Managing Partner, Salomon Brothers

A.C.I. SAMUEL
Director, British Agrochemicals Association

DAVID SCOTT
Chairman and Chief Executive Officer, Allis-Chalmers Corporation

PETER SCOTT
Chairman, Provincial Insurance Company Limited

LORD SEEBOHM
Chairman, Finance for Industry

THE EARL OF SELKIRK
President, Royal Central Asian Society

G.F. SMITH
General Secretary, Union of Construction, Allied Trades and Technicians

*LAUREN K. SOTH
Editor of the Editorial Pages,
Des Moines Register & Tribune

SIR MICHAEL STEWART
Director, The Ditchley Foundation

RALPH I. STRAUS
Director, R.H. Macy & Company, Inc.

JAMES A. SUMMER
Vice Chairman, General Mills Inc.

HAROLD SWEATT
Honorary Chairman of the Board, Honeywell Inc.

SIR ROBERT TAYLOR
Deputy Chairman, Standard and Chartered Banking Group Limited

A.A. THORNBROUGH
President, Massey-Ferguson Limited

LORD TRANMIRE
Thirsk, Yorkshire

SIR MARK TURNER
Deputy Chairman, Kleinwort, Benson, Lonsdale Ltd.

WILLIAM I.M. TURNER
President and Chief Executive Officer, Consolidated Bathurst, Ltd.

HON. JOHN W. TUTHILL
Director General, The Atlantic Institute

CONSTANT M. van VLIERDEN
Executive Vice President, Bank of America, National Trust and Savings Association

R.C. WARREN
Vice President, IBM Corporation, and President, International Operations Division

VISCOUNT WEIR
Chief Executive, The Weir Group Ltd.

WILLIAM W. WINPISINGER
General Vice President, International Association of Machinists and Aerospace Workers

FRANCIS G. WINSPEAR
Edmonton, Alberta

DAVID J. WINTON
Minneapolis, Minnesota

SIR ERNEST WOODROOFE
Formerly Chairman, Unilever Ltd.

SIR MICHAEL WRIGHT
Chairman, Atlantic Trade Study and Director, Guinness Mahon Holdings Ltd.

ARNOLD S. ZANDER†
Professor, University of Wisconsin

---

*See footnote to the Statement.
†Deceased.

Worldwide Financial Implications of Higher Oil Prices

by Sperry Lea

Introduction

In January 1974, the oil-importing countries faced the grim fact that OPEC oil now cost them many times what it had the previous September. Understandably, they felt menaced by sudden challenges, both to their ability to afford their basic oil requirements, and to the institutions, private and public, which normally meet unusual financial needs of governments.

From the present vantage point, these initial alarms may appear somewhat exaggerated, since the financing problems are now considered to have been successfully managed, or at least weathered. Nevertheless, the potency of the underlying threat is not being minimized. One informed commentator now states, "Indeed, the price increases of 1973-74 have probably had a more convulsive effect on the world economy than any other event taking place in the same brief period of time in recorded history."[1] Another judges them to be "leading to the biggest sudden shift of financial resources from one group of nations to another that history has ever witnessed."[2]

The purpose of this report is to supplement the Committee's Policy Statement by offering a concise factual review of the subject. In keeping with the Statement's focus, it concentrates on problems associated with financing oil imports; it deals only in passing with the concurrent issues of dislocation of national economies or security of oil supply—questions that have provoked their own sets of vigorous responses.

The report opens with a preliminary section that clarifies the major terms and concepts. The substance of the topic is discussed

---

1 Alfred Reifman, "U.S. Energy Policy: A Perspective on Major Immediate Issues" (Washington, D.C.: Library of Congress, Congressional Research Service, U.S. Library of Congress, July 24, 1975), p. 4.

2 James P. Grant, "The OPEC Nations: Partners or Competitors?", James W. Howe and the Staff of the Overseas Development Council, *The U.S. and World Development: Agenda for Action 1975* (New York: Praeger, April 1975), p. 135.

in the remaining sections, II and III. The former reviews how the financing problems appeared initially and throughout 1974 and the various developments that have eased them. The final section looks ahead, surveying present preoccupations with three specific issues that are forecast to persist over the rest of this decade. Statistical data and detailed information, some developed expressly for this publication, are collected at its center in 12 Exhibits.

I. What Are We Talking About?

Before turning to the financing problems, past and present, provoked by higher oil prices, it is useful to clarify at the outset some basic concepts and terminology. For ambiguities abound. Indeed, several key words, notably "recycling" and "absorption" prove slippery, being understood by various officials and commentators to encompass quite different ranges of operations. This preliminary section reviews some of the basic terms (italicized when formally introduced) and establishes the vocabulary used thereafter. A special effort is made to identify the many separate modes of recycling from two perspectives, and some new terms are coined to emphasize distinctions that are important in the two main sections of this report.

Dramatis Personae

First, whom are we talking about? The basic transaction is the purchase of oil by the *Oil-Importing Countries* from the 13 members of the Organization of Petroleum Exporting Countries (*OPEC*). While it is sometimes convenient to consider each group collectively, it is usually essential to recognize quite diverse subgroups within each of them.

The Oil-Importing Countries

Virtually all non-OPEC countries outside the Soviet bloc and China are now net *Oil Importers* and rely on OPEC supplies.[1] Exhibit 1 distributes the importing countries among several categories. Of particular relevance to the financing problems is Group B-3, the "most seriously affected" countries (*MSAs*)—sometimes known as *The Fourth World*—within the larger group of Less-Developed Countries (*LDCs*). The official list of the MSAs is given in Exhibit 2.

OPEC Countries

Exhibit 3 presents some basic data on the OPEC countries. These are usefully separated into three Groups, distinctions

---

[1] Canada became a net importer this year, and some Soviet bloc countries will increasingly use OPEC oil in the future. By contrast, several countries expect to cease being net importers in a few years, notably Britain, Norway and Mexico.

between which are rooted in vastly different oil endowment in relation to population. The following figures are derived from Exhibit 3.

Estimated Petroleum Reserves: Barrels per Capita
(OPEC average = 1,440)

Group I	Group II	Group III
Saudi Arabia... 15,640	Iran..........1,830	Nigeria.........320
Kuwait......66,090	Venezuela.....1,160	Indonesia...... 80
U. A. E......127,500	Iraq..........2,810	
Qatar.......65,000	Algeria....... 440	
Libya.......11,130	Equador...... 800	
	Gabon.......2,000	

Particularly relevant to our subject, especially to future prospects, are the high-income Group I countries. These are also called the "low absorbers" since their large per capita oil revenues and reserves place them now, and for some time to come, in the position of earning far more than they can absorb by spending.

The Basic Division of OPEC Oil Revenues

The figures in Exhibit 3 note that in 1973 the OPEC countries received $23 billion in *Oil Revenues* paid by the importing countries. Following the sudden many-fold increase in OPEC prices between October 1973 and January 1974 (Exhibit 5), these oil revenues rose to an estimated $90 billion for 1974, a gain of $67 billion between one year and the next.

As discussed further in the separate Appendix, the OPEC countries use these revenues in two fundamental ways.

Absorption

A growing share of OPEC oil revenues (about one-third in 1974) was used to import goods and services. Payments by OPEC countries for their own imports are called here *Absorption*, more precisely *Direct Absorption*. And, since virtually all these purchases are from the oil-importing countries—specifically those that are industrialized—they offset in part these countries' oil costs in computing their collective current account balances with the OPEC group. Such exports of goods and services by the oil importers also constitute a *Transfer of Real Resources* by them to OPEC countries.

An important variant is the *Indirect Absorption* of OPEC oil revenues, sometimes called *Triangular Absorption*. This occurs when OPEC earnings that are given or loaned to non-oil LDCs are immediately spent on imports where such purchases are attributable solely to such OPEC aid.[2]

Surplus Petrodollars

OPEC oil revenues not offset by OPEC imports (in 1974, roughly $60 billion—the remaining two-thirds[3]) generate *Surplus Petrodollars* or simply *Petrodollars*. ("Petrocurrencies" could well become the more appropriate word.) These funds, also referred to as *OPEC Surpluses*, are initially invested by OPEC countries elsewhere in the world in ways shown in Exhibit 6. For recipient countries, such transactions involve a *Transfer of Financial Claims*.

The remainder of this section concerns the amount of these surplus petrodollars and what is to be done with them.

Measures of OPEC Surpluses

Three concepts relate to the size of OPEC surpluses, the first two having several names since they can be seen from many perspectives.

• The total value of surplus petrodollars generated in a given year is often referred to as the *OPEC Current Account Surplus*, or as its counterpart, the *Importing Countries' Current Account*

---

[2] There is more than one acceptable way to define "absorption" by an OPEC country. It is widely considered—as described here—to be the sum of its gross merchandise and service imports, the former factor being far the more important. So defined, absorption is "direct" or "indirect" depending on whether the OPEC country or a recipient of its aid receives these imports. An alternative definition, used by, among others, J.M. Jefferson (cited below), combines three factors: the OPEC country's merchandise imports; its net service payments (counting as an offset the eventually huge remitted returns on its foreign investments); and Grant/Aid (with the presumption that much of this results in what is called here "Indirect Absorption").

[3] The $60 billion is the roundest number falling between a variety of estimates for the 1974 figure. Note the Bank of England estimate of $56.2 in Exhibit 6, the U.S. Treasury estimate of $58.9 in Exhibit 4, and several others in Exhibit 11-H.

Deficit with OPEC. The use of these funds leads to yet another label, *OPEC's Investable Surpluses*, as used in Exhibit 4.

●The extension of this concept over time is *OPEC's Accumulated Surplus Petrodollars*, the pool that will have built up by a future date, 1980 usually being cited. This aggregate is known also as the *Accumulated OPEC Surplus* (on current account). Its counterpart is the *Accumulated Importing Country Deficit* (on current account with OPEC), also called their *Accumulated Indebtedness* to OPEC. Exhibit 11-I signals the considerable interest in—and disagreement over—the development of OPEC's accumulated surpluses.

●A related concept, typically applied only to an individual importing country, is its incremental *Oil Deficit*, meaning that component of its present current account deficit attributable to the sudden oil price increase of 1973-74. The IMF uses a precise definition of this concept, "the calculated increase in a country's oil import cost," to determine its access to the *Special Oil Facilities*. For reasons given below, the validity of the "oil deficit" as a distinctive concept is waning.

Recycling

Surplus petrodollars begin as financial claims of OPEC countries on the oil companies serving importing countries or on the importing countries themselves. All the various methods of placing and moving these financial claims within and among the importing countries will be called *Recycling*, although, as noted, some people reserve the term for only selected aspects of this process.

There are several ways to categorize recycling, of which two are used here.[4]

... as Distinguished by Locus of Decision

Sections A and B of Exhibit 8 distinguish two modes of recycling surplus petrodollars on the basis of who makes the decisions.

---

[4] A few European commentators refer to OPEC imports ("absorption" here) as "recycling of real assets," but the overwhelming majority reserve the word for movements of financial assets. Others distinguish recycling modes by sequence, producing occasional examples of "tertiary recycling."

Terms and Concepts

●*Primary Recycling* is what the OPEC countries themselves choose to do at the outset with their surplus petrodollars. It encompasses their original placements of funds, all of which take some form of investment in the importing countries or in multilateral institutions expressly organized to relend them. As Exhibit 6 and its recapitulation on page 9 show, about half of OPEC's primary recycling in 1974 was directed to private international financial institutions, mostly in the Eurocurrency markets with some also deposited in the United States and the United Kingdom. This and seven additional modes of primary recycling by OPEC are noted in section A of Exhibit 8.

●*Secondary Recycling* (section B of Exhibit 8) consists of all onward investing by recipients of primary recycling who then act as intermediaries that make their own decisions.[5] Understandably, the principal intermediaries have been the same private financial institutions that receive the lion's share of OPEC's primary recycling (B-1a). Some secondary recycling is done by developed importing countries (B-2), while intermediation by intergovernmental institutions, notably by the IMF (B-3), has become important. *("Recycling" is understood by some to mean only this second mode.)*

. . . as Distinguished by Role

While it is useful to introduce the many forms of recycling on the basis of who decides to use them, it is even more relevant here to divide them according to two quite different roles they can play in helping countries finance their oil deficits.

●For the most part, this task is being accomplished by what will be called *Natural Recycling:* those modes of either primary or secondary recycling that occur spontaneously, the lenders—either an OPEC country or an intermediary—being satisfied to have found a desired mix of safety and return on investment as judged by normal commercial criteria. The major example is on-lending by the private financial institutions, especially by the Euromarkets (B-1a of Exhibit 8).

---

5 OPEC surplus funds placed by primary recycling to the private international institutions or governmental treasuries immediately co-mingle with usually larger pools of funds, losing their distinctive identity as petrodollars. Thus, the concept of secondary recycling can include the on-lending of at least some funds other than those received as petrodollars.

- By contrast is what will be called *Directed Recycling:* those examples of either primary or secondary recycling where funds are loaned on "concessional" rather than commercial terms, or are even granted outright. The specific intent is to help countries which, for one reason or other, cannot meet their oil deficits with "naturally recycled" petrodollars. The main instruments for directed recycling are OPEC aid (A-2b) and loans by the IMF *Special Oil Facility* (B-3a). *("Recycling" is understood by some to mean only "directed recycling.")*
- Indispensable adjuncts to recycling as a response to the oil financing problem are so-called "safety nets" prepared to act as *Lenders of Last Resort.* These *Financial Support Arrangements* deserve to be included as Section C of Exhibit 8, since they bolster the confidence which encourages expanded use of both natural and directed recycling. Besides, in extreme cases they can themselves recycle funds to governments (C-1), to private financial institutions (C-2), or to the IMF (C-3).

Mechanisms now operating to fulfill these three roles—Natural Recycling, Directed Recycling and Lending of Last Resort—are arranged in Exhibit 9 as a spectrum of responses to the initial challenges of financing higher oil costs, to which we now turn.

II. Initial Perceptions and Reactions

During 1974, especially in its early months, the financing problem provoked by higher oil prices was particularly acute. Four related challenges were perceived.

> *(1) The Recycling Problem: Could sufficient funds for borrowing be made available to those countries unable to finance from other sources their balance-of-payments deficits arising from higher oil costs?*

At the beginning of 1974, most oil-importing countries faced the immediate prospects of large and growing current account deficits with OPEC countries. While many could finance at least part of their "oil deficits" by drawing down reserves, earning more abroad and attracting foreign investment, most of them needed to borrow to cover fully their new liabilities.

For the importing countries taken together, no problem existed. OPEC countries were lending them via "primary recycling" of surplus petrodollars the counterpart of their collective current account deficit with OPEC. The task was to assure that these petrodollars could be re-lent via some form of "secondary recycling" from where OPEC countries had placed them to all those countries that needed to borrow them.

As the following 1974 data from Exhibit 6 shows, OPEC concentrated its primary recycling in the developed market economy countries (Group A-1 in Exhibit 1), and within them as deposits in private financial institutions.[1]

Placement of OPEC 1974 surplus in . . .	*$ billion*	*% of total*
Developed Countries: U.K. (lines 1-6), U.S. (7-9), and Other with Eurocurrency institutions (10)	$41	73%
within which *Deposits in Private Financial Institutions*	28.5	51
of which Eurocurrencies (5, 10)	22.8	41
U.S. banks (8)	4.0	7
U.K. Sterling deposits (3)	1.7	3

---

1   This section concentrates on 1974 data. For the first half of 1975, Exhibit 6 shows marked changes in some OPEC placements. Eurocurrency deposits retained roughly the same share but shifted away from London. Meanwhile, net withdrawals were recorded in U.K. Sterling and U.S. bank deposits.

These institutions have in turn been the greatest source of secondary recycling, relending petrodollars on normal terms to countries in need. Exhibit 12 shows the publicized volume of this "natural recycling" during 1974 from its major source, the Eurocurrency markets. Note the extensive borrowing by the more favored "Third World" LDCs—presently over half the total—but their small and apparently declining use by a few "most seriously affected" countries.

Thus, the initial need for additional recycling centered mainly on the MSAs. These countries combine the lowest income levels with large chronic deficits (the greatest being those of India, Bangladesh and Pakistan). The MSAs have rarely used "natural recycling," either finding themselves unable to afford normal terms or being considered insufficiently creditworthy.

(2) The Banking Problem: Could the private international financial institutions fulfill their role as the principal intermediaries, absorbing unprecedented inflows of surplus OPEC oil revenues and relending them without exposure to excessive risk?

Early in 1974, these institutions found themselves virtually the only instruments for secondary recycling of petrodollars, a role confronting them with several traditional banking problems:

- the difficulty of taking in vast amounts of short-term OPEC petrodollars and matching these with far longer-term loans;
- the need to identify, from among many governments now seeking such loans, those that would remain creditworthy despite their suddenly large payments deficits at a time of general economic uncertainty;
- the necessity to maintain given capital-to-deposit ratios in the face of rapidly increasing deposits.

Moreover, these institutions were being pressed to expose themselves to these risks without adequate backstopping by lenders of last resort, a situation emphasized in mid-1974 by several bank failures—although these were for reasons quite unrelated to recycling petrodollars.

(3) The Penalties for Inadequate Response: What might happen in the absence of quick solutions . . .

... to the recycling problem? From the outset loomed the alarming prospect that some importing countries, seeing no way to finance fully their oil deficits, would seek to adjust to them by desperate unilateral actions, for instance:

- trade and payments restrictions or export stimulants, a beggar-thy-neighbor approach inviting a chain reaction of retaliation;
- attempts to reduce their oil deficits by curtailing aggregate economic activity, an approach promoting worldwide economic slowdown, and whose adverse effects on domestic employment could foster extreme political instability at home;
- attempts to make special bilateral deals with particular OPEC countries, a move likely to lead to competitive emulation and eventually less favorable terms for all.

... to the banking problem? If the private financial institutions were to prove unable to cope with recycling tasks whose magnitude and characteristics they were ill-equipped to handle, several serious threats were envisaged:

- to these private financial institutions themselves;
- to the large number of countries relying on them for natural recycling of OPEC surpluses;
- to all oil-importing countries, if those OPEC countries that continued to accumulate significant surpluses—the "low absorbers" from Group I—now considered it a wiser "investment" to leave some of their oil in the ground.

(4) The Dimensions of the Financing Challenge: To what size would the pool of accumulated OPEC surpluses grow, and for how long would it exist?

Underlying the three other concerns within the oil-importing countries were the volume and duration of their accumulated indebtedness to OPEC. Indeed, during 1974 the key question was whether there was any end in sight. Largely responsible for this view was the first widely publicized estimate, that of the World Bank in July 1974.[2] This projected OPEC surpluses to reach over $650 billion in current value in 1980, and $1.2 trillion by 1985

---

2 This estimate appeared in a World Bank study that was not a public document, a fact that probably enhanced the publicity given it.

and, more important, to be then still growing by over $100 billion per year.

This projection, which dominated perceptions of the financing challenges through the remainder of 1974, was replaced in 1975 by a number of more moderate estimates, major examples of which are given, together with their basic assumptions, in Exhibit 11.

Improved Recycling

To these challenges, the most direct policy response by both the importing and the OPEC countries has been to expand considerably the number and variety of mechanisms for recycling surplus petrodollars to where they were needed, and to provide several "safety nets" to reinforce both these mechanisms and the countries they serve.

Exhibit 9 summarizes the present range of techniques to affect or support recycling to governments, grouping them in the three categories developed above.

Natural Recycling

●The *Private International Financial Institutions* have remained the principal instrument for financing national oil deficits on normal terms. Though these pre-dated the oil crisis, their suddenly expanded role required both supplementing by mechanisms expressly designed to serve the MSAs, and backstopping by lenders of last resort. New measures to these two ends are discussed immediately below.

●Meanwhile, some *OPEC Lending to Developed Countries* on commercial terms took place in 1974: for example, Iranian loans to Britain and other European countries, the Saudi Arabian purchase of nonmarketable U.S. securities, and OPEC placements of funds with two provincial hydroelectric utilities in Canada.

Directed Recycling

The major innovations fall under this category. They were designed to meet both the recycling problem, by serving needy countries that could not borrow on normal terms, and the banking problem, by supplementing the private financial institutions where these are not equipped to operate. There are two major sources of directed recycling: the IMF's two successive Oil Facilities and related Subsidy Account, and various forms of OPEC aid.

• The *First IMF Oil Facility* (June 1974-June 1975) was designed to recycle funds, mainly from OPEC countries, to IMF members irrespective of their stage of development but having oil-induced deficits that needed financing.[3] Exhibit 10 shows that during its 12-month duration, the First Facility received commitments for 3.0 billion SDRs ($3.8 billion) from 7 OPEC and 2 other countries, and recycled 2.6 billion SDRs ($3.2 billion) to a total of 40 countries. As derived from Exhibit 10, the distribution of lenders and borrowers to the First Facility is as follows:

	LENDERS		BORROWERS			
			Developed Countries		LDCs	
	OPEC	Others	Industrial	Other	3rd World	MSAs
No. of Countries	7	2	1	7	9	23
% of total value	87%	13%	26%	31%	16%	27%

• The *Second IMF Oil Facility* was negotiated early in 1975.[4] New commitments to lend were obtained and borrowing began that June. Exhibit 10 includes details of its operations through September 1975, showing a widening lending role for non-OPEC countries with strong reserve positions. As just noted, the Second Facility alters the conditions for borrowing to give less emphasis now to an applicant's "oil deficit," and more to the size of its regular IMF quota, a move signaling the probable replacement of

---

3   One determinant of a country's access to both Oil Facilities has been "the calculated increase in its oil import cost." This is the product of the volume of its net imports of petroleum and petroleum products in 1972 or 1973 (whichever was higher) by $U.S. 7.50 per barrel (the 1973-74 increase in the OPEC price). For the First Facility, an applicant country's access could not exceed 100 percent of its "oil deficit" thus calculated or 75 percent of its IMF quota, whichever was smaller. For the Second Facility, these ratios have been switched to 85 percent of its "oil deficit" and 125 percent of its IMF quota.

4   During its gestation period, the renewed Oil Facility was frequently named for either of two proponents: "Witteveen II" in honor of the Managing Director of the IMF, or the "Healey Plan," recognizing the initiative by Denis Healey, the British Chancellor of the Exchequer. The negotiations alluded to involved compromises necessary to achieve simultaneous IMF acceptance in January 1975 of both the renewed Oil Facility and the so-called "Kissinger-Simon Plan" for an OECD Financial Support Fund, discussed below.

Special Oil Facilities by more general purpose IMF lending operations.

● Recognizing that the 7½+ percent interest rates of the Oil Facility are too burdensome for the MSAs, the IMF launched in the summer of 1975 a special *Subsidy Account*. Contributions are to come from IMF member countries in a position to make them.[5] Payments will go to MSAs on the official UN list that have drawn from the Second Oil Facility, with the effect of subsidizing in large part their interest payments for such borrowings.

● Supplementing the major role of OPEC countries in lending to the Oil Facilities is *OPEC Bilateral Aid to LDCs*. This provides directed primary recycling in many forms: concessional loans, grants, investments, and sales of oil on easy terms.[6] Such aid is not restricted to meeting the recipient's "oil deficits," though it serves this function among others. The volume of bilateral aid increased sharply in 1974 to an estimated $2.0 billion in actual disbursements, with commitments for future aid of over $7 billion. Exhibit 7-A shows the principal donors to be Saudi Arabia, Iran and Kuwait. Among the recipients of bilateral aid in 1974, Exhibit 7-B identifies 16 from the MSA roster, accounting for about 90 percent of disbursements and 95 percent of commitments. A recapitulation of this data, showing how OPEC's bilateral aid was distributed in 1974, reveals, however, that emphasis has been sharply focused (see tabulation on page 15).

Last Resort Lending

Exhibit 9 ends by listing three financial support arrangements that backstop developed governments by acting as lenders of last resort on regular commercial terms:

---

5   By mid-August 1975, 18 IMF countries had expressed a willingness to contribute a total of 149 million SDRs to the subsidy account, with additional contributions expected.

6   Any attempt to describe OPEC bilateral aid accurately and fairly confronts several obstacles. For one thing, the basic data is sparse, the OECD estimates in Exhibit 7 being the only comprehensive figures by country available. Internal estimates of the World Bank and the IMF appear to show considerably higher levels of present OPEC aid, and there are news stories of $500 million in OPEC aid to Pakistan alone in the year ending June 30, 1975. Then, as with all foreign aid programs, OPEC's cover a wide spectrum between doing good and doing well, with its loans to the IMF and World Bank considered by some to be among the soundest of investments under present conditions, especially in terms of safety. Finally, the rise of "Indirect Absorption" complicates the picture.

Exhibit 1: A Classification of the (Net) Oil-Importing Countries

A. Developed Countries	Major Examples
1. *Market Economies*	All members of the OECD, Canada just now included. (Britain and Norway expected off the list as net exporters by 1980.)
2. *Communist*	Most communist countries except the U.S.S.R., and China. But communist oil importers are the only such countries not basically dependent on OPEC oil.

B. Less-Developed Countries (LDCs)*

1. *Relatively Favorably Endowed*

    a. ... with oil — Mexico, Bolivia, Columbia, and Peru (of which Mexico is expected soon to be a net exporter and off this list). Other countries, e.g., India and Greece, are considered to have potentially significant oil production.

    b. ... with other marketable raw materials — Includes many LDCs in Latin America and Eastern Africa plus Malaysia (generally excludes countries in West Africa, South and Southeast Asia).

2. *Linked Economically with Developed Countries*

    a. ... through geographical closeness, resulting in tourism, workers' remittances, and exports of agricultural perishables — Numerous countries of the Mediterranean rim, in the Caribbean, Central America, and close to Japan.

    b. ... through processing certain goods for the world economy — South Korea, Taiwan, Hong Kong, Singapore, Mexico.

3. *"Most Seriously Affected" (MSAs)*

    ... by higher oil prices and other aspects of present world economic situation — The UN presently lists 41 MSAs, so designated because of per capita incomes below $400 and projected payments deficits at least 5% of imports. (See list in Exhibit 2.)

*A number of LDCs enjoy inclusion in several categories of B-1 and B-2, notably Mexico.

Source: This classification of LDCs roughly follows that of James Grant in "Energy Shock and the Development Prospect" in James Howe, *The U.S. and the Developing World: Agenda for Action, 1974* (New York: Praeger for the Overseas Development Council, 1974).

Exhibit 2: The "Most Seriously Affected Countries" (MSAs)

U.N. List at end of 1974

Country	Per capita GNP* (U.S. dollars)
Rwanda	$ 60
Bangladesh	70
Mali	70
Somalia	70
Upper Volta	70
Chad	80
Ethiopia	80
Guinea	90
Yemen Arab Republic	90
Dahomey	100
Lesotho	100
Niger	100
Sri Lanka	100
India	110
Tanzania	110
Haiti	120
Laos	120
Yemen, People's Democratic Republic	120
Khmer Republic	130
Pakistan	130
Sudan	130
Malagasy Republic	140
Central African Republic	150
Kenya	160
Mauritania	170
Cameroon	200
Sierra Leone	200
Ghana	250
Senegal	250
Honduras	300
El Salvador	320
Ivory Coast	330
Guyana	390

Added to List in April 1975

Burundi	60
Afghanistan	80
Burma	80
Uganda	130
Western Samoa	140
Cape Verde Islands	180
Egypt	220
Mozambique	280

*As of 1971.

The MSAs (B-3 of Exhibit 1) are those countries with a per capita income below $400 and which had a projected balance-of-payments deficit for 1974 and 1975 not smaller than 5% of imports.

Source: *IMF Review*, April 28, 1975.

Exhibit 3: Some Basic Data on OPEC Countries

	Estimated Oil Revenues 1973	Estimated Oil Revenues 1974[b]	Non-Oil Exports, 1973	Imports, 1973[c]	Estimated Per Capita GNP[a] 1973	Estimated Per Capita GNP[a] 1974	Population, mid-1975	International Reserves, January 1975	Crude Production, 1973	Petroleum Reserves at 1973 Production Rate	Estimated Petroleum Reserves
	($ millions)	($ millions)	($ millions)	($ millions)	($)	($)	(millions)	($ millions)	(million barrels/day)	(years)	(billion barrels)
Group I: High Income/Low Absorption											
Saudi Arabia	5,100	20,000		1,993	980	2,900	9.0	14,285[e]	7.7	51	140.8
Kuwait	1,900	7,000		1,042	4,100	8,500	1.1	1,654	3.1	66	72.7
United Arab Emirates	900[f]	4,100[f]	577[d]	800	9,000	21,000	0.2	n.a.	1.5	45	25.5
Qatar	400	1,600		170	3,300	12,000	0.1	n.a.	0.5	31	6.5
Libya	2,300	7,600	106	1,723	3,000	5,800	2.3	3,523	2.2	32	25.6
Group II: Middle Income/High Absorption											
Iran	4,100	17,400	864	3,370	520	940	32.9	8,513	5.9	28	60.2
Venezuela	2,800	10,600	1,232	2,813	1,150	1,850	12.2	6,191	3.5	11	14.2
Iraq	1,500	6,800	22	899	430	930	11.1	3,273[e]	2.0	44	31.2
Algeria	900	3,700	270	2,338	350	530	16.8	1,497	1.0	20	7.4
Ecuador	100	800	n.a.	532	320	420	7.1	336	0.2	78	5.7
Gabon[g]	100	400	n.a.	160	900	1,540	0.5	47[h]	0.2	n.a.	1.0
Group III: Low Income/High Absorption											
Nigeria	2,000	7,000	287	1,874	150	230	62.9	5,981	2.0	27	19.9
Indonesia	900	3,000	1,923	2,347	80	100	136.0	1,624	1.3	22	10.8
Total OPEC	23,000	90,000	5,281	20,061			292.2	46,924	30.9		420.5

[a] The estimates for per capita GNP of some OPEC countries vary widely among sources, sometimes by 50 per cent or more. The World Bank's estimates for 1974 are not yet available; its estimates for 1973 are the following: Saudi Arabia, 690; Kuwait, 4,354; United Arab Emirates, n.a.; Qatar, 3,447; Libya, 2,007; Iran, 585; Venezuela, 1,303; Iraq, 443; Algeria, 432; Ecuador, 392; Gabon, 988; Nigeria, 152; Indonesia, 102.

[b] These are highly tentative estimates. The most recent guess is that OPEC revenues, when calculated on an accrual basis rather than on payment actually received, reached some $100 billion in 1974.

[c] The International Monetary Fund estimates the 1974 import bill for the 12 OPEC members plus Gabon, Bahrain, Brunei, Oman, and Trinidad and Tobago at $36 billion, a sharp rise from their $21.3 billion import bill for 1973.

[d] Figure is the total figure for Saudi Arabia, Kuwait, Abu Dhabi (a member of the United Arab Emirates), and Qatar.

[e] December 1974 figure.

[f] Figure is for Abu Dhabi, a member of the U.A.E.

[g] An Associate Member of OPEC.

[h] August 1974 figure.

SOURCES: Estimates for oil revenues and non-oil exports, and most figures for per capita GNP, are from the Report by the Chairman of the Development Assistance Committee, *Development Co-operation, 1974 Review* (Paris: OECD, 1974), p. 44. Per capita GNP figures for U.A.E. and Libya are from *The Economist*, February 15, 1975, p. 72. Figures for imports and international reserves are taken from International Monetary Fund, *International Financial Statistics*, Vol. 28, No. 3, March 1975. Population figures are from Population Reference Bureau, "1975 World Population Data Sheet." Figures for crude production, years of petroleum reserves, and petroleum reserves are from *Business Week*, January 13, 1975, p. 80, with the exception of Gabon, whose figures for crude production and petroleum reserves are from The Royal Dutch Shell Group of Companies, *Information Handbook 1974-5* (London: Shell International Petroleum Company Limited, 1974), pp. 65 and 67.

As reproduced in James W. Howe and the staff of the Overseas Development Council, *The U.S. and the World Development: Agenda for Action 1975* (New York: Praeger, for the Overseas Development Council, April 1975), pp. 240-41. Excerpted and reprinted by permission of Praeger Publishing Co.

Exhibit 4: Estimated OPEC Investible Surplus for 1975, with Summary Data for 1973 and 1974
($ Billions)

	1975						
	Oil Exports (Gov't. Take)	Non-Oil Exports	Imports F.O.B.	Services & Private Transfers	Investible Surplus		
					1975	*1974*	*1973*
Group I	48.1	0.6	-14.5	0	34.3	*36.3*	*4.4*
Saudi Arabia	26.7	—	-5.7	-0.1	21.0	*20.8*	*3.1*
Kuwait	7.9	0.5	-2.1	0.8	7.1	*7.3*	*1.5*
U. A. E.	6.5	—	-2.2	-0.1	4.2	*4.4*	*0.3*
Qatar	1.8	—	-0.4	-0.1	1.3	*1.3*	*0.1*
Libya	5.2	0.1	-4.1	-0.5	0.7	*2.5*	*-0.6*
Group II	39.9	2.5	-30.3	-2.3	9.8	*17.2*	*0.7*
Iran	19.9	1.0	-10.6	-0.7	9.6	*10.7*	*1.1*
Venezuela	8.3	0.5	-6.5	-0.5	1.8	*4.0*	*-0.1*
Iraq	7.6	0.2	-6.6	-0.7	0.5	*2.0*	*0.5*
Algeria	3.7	0.3	-5.7	-0.3	-2.0	*0.4*	*-0.8*
Equador	0.4	0.5	-0.9	-0.1	-0.1	*0.1*	—
Group III	10.4	3.3	-9.8	-2.3	1.6	*5.4*	*-0.1*
Nigeria	6.7	0.9	-5.1	-0.6	1.9	*5.2*	*0.3*
Indonesia	3.7	2.4	-4.7	-1.7	-0.3	*0.2*	*-0.4*
OPEC TOTAL 1975	98.1	6.6	-54.5	-4.5	45.7	*58.9*	*4.9*
1974	*94.9*	*5.7*	*-36.9*	*-4.8*	*58.9*		
1973	*25.2*	*4.3*	*-20.3*	*-4.3*	*4.9*		

— = less than $50 million.

Columns may not add due to rounding.

Source: Derived from tables prepared by the U.S. Treasury Department in August 1975 and annexed to "The Absorptive Capacity of the OPEC Countries," released September 5, 1975.

Exhibit 5: OPEC Oil Prices, 1970-74

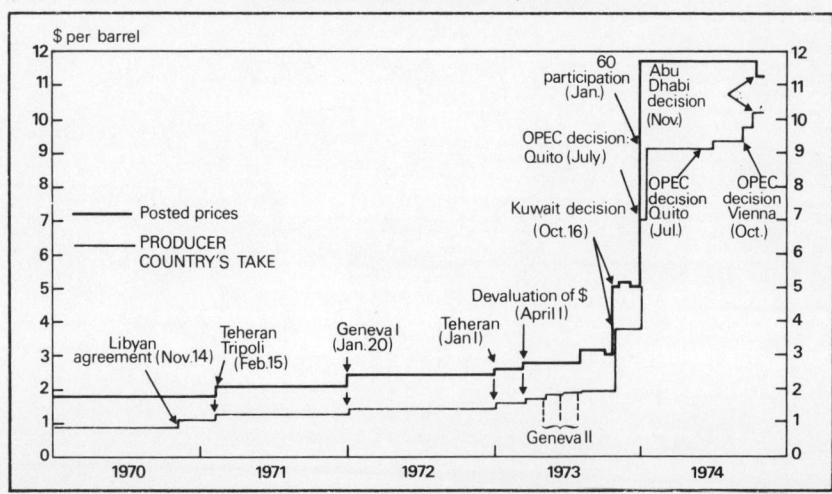

Source: *The Financial Times* (London) as published in *Conjuncture* (Paris: Société Générale, January 1975).

Exhibit 6: Primary Recycling: Estimated Deployment of OPEC Surpluses, 1974 and First Half of 1975
($ Billions)

	1974						1975	
	I	II	III	IV	Total		1st Half	
	$	$	$	$	$	%	$	%
United Kingdom								
1. British government stocks	0.4	0.1	0.2	0.2	0.9	1.6	0.3	1.8
2. Treasury bills	0.4	0.7	0.7	0.9	2.7	4.8	0.2	1.2
3. Sterling deposits	-0.1	0.7	1.1	—	1.7	3.0	-0.1	-0.1
4. Other sterling investments (a)	0.1	0.2	0.3	0.1	0.7	1.3	0.1	0.1
5. Foreign currency deposits	2.5	4.5	3.4	3.4	13.8	24.6	2.0	12.0
6. Other foreign currency borrowing	—	0.5	0.3	0.4	1.2	2.1	0.2	1.2
	3.3	6.7	6.0	5.0	21.0	37.4	2.7	16.2
United States								
7. Government and agency securities	0.5	1.4	2.3	1.8	6.0	10.7	1.3	7.8
8. Bank deposits	0.6	0.8	2.3	0.3	4.0	7.1	-0.1	-0.1
9. Other (a)	—	0.1	0.4	0.5	1.0	1.8	1.0	6.0
	1.1	2.3	5.0	2.6	11.0	19.6	2.2	13.7
Other Countries								
10. Foreign currency deposits	1.5	3.5	1.5	2.5	9.0	16.0	5.0	29.9
11. Special bilateral facilities and other investments (a) (b)	1.1	2.5	3.0	5.0	11.6	20.6	5.3	31.7
	2.6	6.0	4.5	7.5	20.6	36.6	10.3	61.7
International organizations	—	0.5	0.8	2.3	3.6	6.4	1.5	9.0
TOTAL	7.0	15.5	16.3	17.4	56.2	100.0	16.7	100.0

(a) Includes holdings of equities and property. (b) Includes loans to less-developed countries.

Source: Bank of England *Quarterly Bulletin*, June and September 1975.

Exhibit 7: OPEC Aid: Commitments and Disbursements in 1974

A. OPEC Donors of Bilateral and Multilateral Aid

($ Millions and Percentages)

	Commitments					Disbursements[a]				
	Bilateral	Multi-lateral	Total	As Percentage of: Oil Revenues[a]	As Percentage of: GNP[a]	Bilateral	Multi-lateral	Total	As Percentage of: Oil Revenues	As Percentage of: GNP
	($ millions)					($ millions)				
Algeria	31	108	139	3.8	1.7	15	45	60	1.1	0.7
Iran	2,802	173	2,975	17.1	10.1	600	2	602	3.4	2.0
Iraq	222	58	280	4.1	2.9	80	35	115	1.7	1.2
Kuwait	957	384	1,341	19.1	15.8	460	70	530	7.6	6.2
Libya	178	241	419	5.5	3.4	45	60	105	1.4	0.9
Nigeria	1	16	17	0.2	0.2	1	15	16	0.2	0.1
Qatar	97	60	157	9.8	7.4	40	20	60	3.7	2.8
Saudi Arabia	2,568	453	3,021	15.1	3.4	650	115	765	3.8	3.4
United Arab Emirates	296	182	478	11.6	10.0	120	45	165	4.0	3.5
Venezuela	20	726	746	7.0	3.6	20	165	185	1.7	0.9
Total	7,172	2,401	9,573	12.1	8.2	2,031	572	2,603	3.3	2.2

[a] Estimate.

NOTE: Figures are subject to revision. Some of the disbursement estimates are highly tentative, as are all the comparisons of commitments and disbursements with oil revenues and GNP.

SOURCE: Organisation for Economic Co-operation and Development, Development Assistance Directorate, "Flow of Resources from OPEC Members to Developing Countries," Document No. DD-403, December 6, 1974, Table 2, p. 8.

Exhibit 7: OPEC Aid: Commitments and Disbursements in 1974 (cont'd.)

B. Recipients of OPEC Bilateral Aid

($ Millions and Percentages)

	Commitments		Disbursements[a]	
	($ millions)	(as percentage of total)	($ millions)	(as percentage of total)
Arab Countries				
Egypt*	3,121[a]	43.5[a]	765	37.7
Syria	1,003[a]	14.0[a]	325	16.0
Jordan	185[a]	2.6[a]	140	6.9
Mauritania*	153[a]	2.1[a]	25	1.2
Sudan*	107	1.5	70	3.4
Somalia*	82	1.1	25	1.2
Morocco	80	1.1	25	1.2
Tunisia	54	0.8	15	0.7
Bahrain	21	0.3	10	0.5
Yemen, Arab Rep.*	19	0.3	15	0.7
Yemen, People's Dem. Rep.*	12	0.2	10	0.5
Other	8	0.1	9	0.4
Africa				
Malagasy Republic*	114[a]	1.6[a]	—	—
Guinea*	16	0.2	5	0.2
Uganda*	12	0.2	2	0.1
Senegal*	11	0.1	5	0.2
Other	79	1.1	46	2.3
Asia				
Pakistan*	957[a]	13.3[a]	355	17.5
India*	945	13.2	75	3.7
Sri Lanka*	86	1.2	35	1.7
Bangladesh*	82	1.1	50	2.5
Latin America				
Guyana*	15	0.2	15	0.7
Honduras*	5	0.1	5	0.2
Europe				
Malta	5	0.1	5	0.2
Total	7,172[a]	100.0	2,031	100.0

[a]Estimate. *An MSA roster, Exhibit 2.

NOTE: Figures are subject to revision. Breakdown between Egypt, Syria, and Jordan is tentative. Commitment figures for Pakistan include $200 million by Saudi Arabia and Kuwait which were not made public but are believed to have been made.

SOURCE: Organisation for Economic Co-operation and Development, Development Assistance Directorate, "Flow of Resources from OPEC Members to Developing Countries," Document No. DD-403, December 6, 1974, Table 4, p. 10.

As reproduced in James W. Howe and the staff of the Overseas Development Council, *The U.S. and World Development: Agenda for Action 1975* (New York: Praeger, for the Overseas Development Council, April 1975), pp. 262-63. Reprinted with the permission of Praeger.

Exhibit 8: Modes of Recycling Surplus Petrodollars

Phases and Methods	Examples and Comments
A. PRIMARY RECYCLING: Initial Placements by OPEC Countries in Importing Countries	
1. OPEC Governments to Private Sectors in Developed Countries	
a. Investing in International Financial Institutions *(the major mode of Primary Recycling to date)*	OPEC oil sales involve credit transfers to account of supplying governments, which concentrate these credits in deposits—generally short term—with major international banking houses in Europe and the U.S., with assets denominated primarily in dollars and pounds. *1974: to Eurocurrency Markets, $22.8 billion; to U.S. banks, $40 billion; to sterling deposits in U.K., $1.7 billion (Exhibit 6).* See B-1 below for on-lending of these assets.
b. Portfolio Investments	Increasingly, OPEC money is placed in U.S. and U.K. stocks and bonds, but is difficult to trace.
c. Direct Investments	Examples include Iran in Krupp, the European Uranium Consortium, and possibly Pan Am; Abu Dhabi in London real estate; Kuwait in British and European real estate and South Carolina resort property. Actual volumes believed well below portfolio investments, though far more publicized.
2. OPEC Governments to Foreign Governments (Developed and LDCs)	
a. Purchases of Developed Countries' Securities, and Official Loans	U.K. and U.S. government securities. *1974: $6 and $3.6 billion respectively (Exhibit 6).* Also, lending to Japan, Italy, Canadian provinces, etc.
b. Aid to LDCs *(a mode of Directed Recycling)*	Forms include grants, soft loans and concessional oil sales. Aid can be bilateral or channeled via co-religious or regional organizations (e.g., Arab Groups, Inter-American bank). *1974: OPEC aid disbursements $2.6 billion of which $2.0 billion bilateral, with bilateral commitments over $7 billion (Exhibit 7).*
c. Loans to IMF "Special Oil Facilities" *(Directed Recycling)*	OPEC commitment to First Facility: *SDR 2.65 billion (Exhibit 10A);* Pledges for Second Facility: *over SDR 2 billion thus far.*
3. Private OPEC Citizens to Private Sectors in Developed Countries	
a. Portfolio Investments	Presumed being made but not identifiable.
b. Direct Investments	Bank of the Commonwealth, Detroit: control bought by Ghaith Pharaon (Saudi Arabian). Investments worldwide by Adnan Khashoggi (Saudi Arabian), etc.

1. ... by the Private International Financial Institutions
   a. to Consuming Country Governments *(largest mode of Secondary and Natural Recycling)*

   Thus far, the largest single source of governmental borrowing to finance "oil deficits." *1974: Publicized Eurocurrency credits to World Bank countries, $26.5 billion (Exhibit 12).*

   b. to Productive Enterprise

   A typical function of the banking system.

2. ... by Consuming Countries
   a. to Other Developed Countries

   German loan to Italy.
   Canada and the Netherlands committed SDR 388 million to the 1974 Facility *(parallel to OPEC's contribution A-2c above).* Seven developed countries with a strong reserve position have pledged SDR 860 million thus far to Second Facility *(Exhibit 10A).*

   b. to IMF "Oil Facilities" *(Directed Recycling)*

3. ... by Intergovernmental Agencies
   a. Loans by IMF Oil Facility *(Directed Recycling)*

   The First Facility recycled SDR 2.6 billion to 40 countries *(Exhibit 10B),* Second Facility in operation will be supplemented by special Subsidy Account to reduce the interest rates to MSAs.

C. LENDING OF LAST RESORT: By "Financial Support Arrangements"

1. ... to Governments
   a. European Community Loan Plan

   Community has authority to borrow up to $3 billion from OPEC countries (Saudi Arabia and Kuwait agreed to provide about $1 billion thus far) to recycle, as a last resort, to members with oil-induced deficits.

   b. OECD Financial Support Fund (originally, Kissinger Plan)

   Up to $25 billion, pooled by OECD members, will be available for recycling to member in need when all other means of financing are exhausted and the applicant considered has attempted serious measures to conserve energy and develop new resources; a "safety net" of last resort.

   c. Swap Arrangements

   These short-term agreements among governments continue in existence.

2. ... to International Financial Markets
   a. European Central Bank Agreement

   European central banks agreed in June 1974 to act as lenders of last resort to Euro-banks (similar to Federal Reserve agreement with U.S. banks).

3. ... to the IMF
   a. General Agreement to Borrow

   The Group of Ten (industrialized countries) renewed and expanded the General Agreement to Borrow (GAB, by which they would lend over SDR 5 billion in their currencies to the IMF.

Exhibit 9: The Spectrum of Recycling to Governments

		Start of Operation	Applicability	
	Mechanisms (Exhibit 8 references)		Developed Countries	Less-Developed Countries
NATURAL RECYCLING for lending at commercial rates	Borrowing from Private International Financial Institutions on commercial terms (Secondary Recycling: B-1a)	Pre-dates oil price rise	All that are creditworthy (thus only slight possibility for exceptions)	All that are creditworthy and can afford interest rates (thus generally excludes MSAs)
	Borrowing from OPEC countries on commercial terms (Primary Recycling: A-2a)	Since early 1974*	Same as above	Not generally applicable
DIRECTED RECYCLING from lending at concessional rates offered specifically to meet oil-induced deficits	Borrowing on concessional terms from OPEC countries or group established by them (Primary Recycling: A-2b)	Since early 1974	Not applicable	Offered thus far mostly to LDCs close to major OPEC countries in terms of region or geography
	Borrowing from the IMF Oil Facilities at 7%-8% interest up to 7 years in amounts limited by an applicant's oil-induced deficit and IMF quota (Secondary Recycling: B-3a)	First Facility operated June '74–June '75; Second Facility since then	All IMF members that meet the conditions for access—including other efforts to solve problems and have shunned restrictive measures	All IMF members meeting conditions listed at left
	Drawing from IMF special "Subsidy Account" to reduce the burden of interest payable to the Second Facility (Secondary Recycling: B-3a)	Adopted in second half of 1975	Not applicable	Applicable only to MSAs drawing from Second Facility
LAST RESORT LENDING by Financial Support Arrangements	Utilising regular Short-Term Swap lines (C-1c)	Pre-dates oil price rise	Applicable to central banks of most developed countries	Not applicable
	Drawing from Community Loan Fund at commercial rates (C-1a)	Available since Feb. 1975	EC countries only	Not applicable
	Drawing from OECD Financial Support Fund (C-1b)	Expected to be available by end of 1975	OECD countries only	Not applicable

*Some lending of this kind by Kuwait before 1974.

Exhibit 10: The Operations of the IMF Oil Facilities (through September 1975)[1]
(Millions of SDRs)[2]

A. LENDERS

	1st Facility		2nd Facility	
OPEC COUNTRIES	2,650	87%	2,010	70%
Saudi Arabia	1,000		1,000	
Iran	580		410	
Venezuela	450		200	
Kuwait	400		200	
Abu Dhabi	100			
Nigeria	100		200	
Oman	20			
OTHER COUNTRIES	388	13%	860	30%
Canada	238		200	
Netherlands	150		300	
Germany			150	
Switzerland			100	
Belgium			50	
Austria			50	
Norway			50	
Trinidad & Tobago			10	
TOTAL CONTRIBUTIONS	3,038	100%	2,870	100%
Carried over from 1st Oil Facility			450	
TOTAL FUNDS AVAILABLE	3,038		3,320	

B. BORROWERS

	1st Facility		2nd Facility	
INDUSTRIAL COUNTRIES	675.0	26%	780.2	51%
Italy	675.0		780.2	
OTHER DEVELOPED COUNTRIES	802.7	31%	237.6	16%
Cyprus	8.1			
Greece	103.5		51.8	
Iceland	17.2		8.4	
New Zealand	109.3		49.5	
Spain	296.2			
Turkey	113.2		56.6	
Yugoslavia	155.2			
Finland			71.3	
LDCs: "THIRD WORLD"	414.0	16%	188.9	12%
Chile	118.5			
Costa Rica	18.8		12.0	
Fiji	.3			
Israel	62.0			
Korea	100.0		58.4	
Nicaragua	15.5			
Panama	7.3			
Uruguay	46.6		21.6	
Zaire	45.0			
Philippines			96.9	
LDCs: MSAs "FOURTH WORLD"	691.0	27%	316.1	21%
Bangladesh	51.5			
Burundi	1.2			
Cameroon	4.6			
Cent. Af. Rep.	3.3		1.7	
Chad	2.2			
El Salvador	17.9			
Ghana	38.6			
Guinea	3.5			
Haiti	4.8		1.6	
Honduras	16.8			
India	200.0		201.3	
Ivory Coast	11.2			
Kenya	36.0			
Malagasy Rep.	14.3			
Mali	5.0			
Pakistan	125.0		76.4	
Senegal	15.5		9.9	
Sierra Leone	4.9			
Sri Lanka	43.5			
Sudan	28.7			
Tanzania	31.5		20.6	
Uganda	19.2			
Yemen, P.D.R.	11.8		4.6	
TOTAL BORROWINGS	2,582.7	100%	1,522.8	100%

[1] Data for the First Facility cover the 12 months from mid-June 1974 to mid-June 1975, and for the Second Facility for the 3½ months from then through September 1975.

[2] In mid-August 1975 one SDR was worth $US 1.19; $Can 1.23; and £0.56.

Source: Compiled from IMF announcements.

Exhibit 11: OPEC Current Account Estimates
(Billions of Current US $)

		1974	1975	1976	1977	1978	1979	1980
A.	**OPEC oil exports (MMB/D)**							
	Enders		27.0		31.0			25.0
	W. J. Levy	29.6	26.5	30.0	31.5	32.5	32.5	31.5
	Morgan Guaranty a/	29.6						29.6
	Irving Trust b/	30.5	28.6	27.3	24.7	24.7	24.7	24.7
	Citibank	30.0	26.0	27.0	28.0	29.0	30.0	31.0
	CIA	29.2	26.1	27.7	29.3			23.6
B.	**Per barrel revenues ($/B)**							
	W. J. Levy	9.72	10.00	11.20	12.00	12.80	13.70	14.65
	Morgan Guaranty a/	9.72						13.25
	Irving Trust c/							
	- Prices rise	9.40	11.00	12.00	12.87	13.74	14.70	15.72
	- Prices break	9.40	11.00	12.00	10.00	9.00	8.00	7.00
	Citibank c/	11.40	11.30	11.80	11.20	10.70	9.90	9.10
	CIA	10.68	10.89	11.73	12.35			
C.	**OPEC total oil revenues**							
	IBRD		105.1					185.1
	Enders	d/	d/		148.8			
	W. J. Levy	95.0	93.0	123.0	138.0	152.0	162.0	168.0
	Morgan Guaranty	105.0	110.0	119.0	125.0	128.0	135.0	143.0
	Irving Trust e/							
	- Prices rise	108.0	119.0	120.0	116.0	124.0	132.0	141.0
	- Prices break	108.0	119.0	120.0	90.0	81.0	72.0	63.0
	Citibank	126.0	107.0	116.0	114.0	113.0	108.0	103.0
	CIA	113.9	103.8	118.7	131.9			
D.	**Non-oil exports f/**							
	IBRD		6.1					21.0
	W. J. Levy	7.0	8.0	10.0	12.0	14.0	16.0	19.0
	Morgan Guaranty	7.0	8.0	10.0	12.0	14.0	16.0	19.0
	Irving Trust	12.0	14.0	17.0	21.0	25.0	30.0	36.0
	CIA	5.9	7.1	8.5	10.5	12.9		
E.	**Imports (Goods and Services)**							
	IBRD (Goods only)		58.4					160.6
	W. J. Levy	45.0	58.0	71.0	88.0	108.0	133.0	164.0
	Morgan Guaranty	50.0	65.0	83.0	108.0	138.0	177.0	227.0
	Irving Trust g/	57.0	76.0	98.0	119.0	144.0	176.0	215.0
	Citibank h/	65.0	79.0	92.0	105.0	118.0	131.0	146.0
	CIA	48.9	70.8	89.5	108.7			
	CIA (goods only)	35.5	54.0	68.7	84.4			
F.	**Investment Income**							
	W. J. Levy	3.0	7.0	11.0	16.0	21.0	26.0	30.0
	Morgan Guaranty	3.0	8.0	13.0	16.0	19.0	19.0	16.0
	Irving Trust							
	- Prices rise	2.0	7.0	12.0	16.0	19.0	21.0	21.0
	- Prices break	2.0	7.0	12.0	16.0	17.0	15.0	10.0
	CIA	4.1	7.7	11.2	14.8			
G.	**Grant Aid**							
	W. J. Levy	2.0	3.0	4.0	5.0	6.0	6.0	6.0
	Morgan Guaranty	2.0	3.0	3.0	3.0	3.0	3.0	3.0
	Citibank i/	4.0	6.0	5.0	3.0	3.0	3.0	3.0
	CIA	4.2	2.7	2.2	2.1			

Exhibit 11: OPEC Current Account Estimates (cont'd.)
(Billions of Current US $)

	1974	1975	1976	1977	1978	1979	1980
H. **OPEC Current Account** (cash basis)							
IBRD		47.2					47.9
Enders		56.0	61.0	64.7			
W. J. Levy	58.0	47.0	69.0	73.0	73.0	65.0	47.0
Morgan Guaranty	63.0	57.0	54.0	40.0	17.0	-13.0	-56.0
Irving Trust							
- Prices rise	65.0	64.0	51.0	34.0	24.0	7.0	-17.0
- Prices break	65.0	64.0	51.0	8.0	-21.0	-59.0	-106.0
Citibank	66.0	36.0	37.0	30.0	19.0	8.0	-7.0
CIA	56.8	45.1	46.7	46.3			
I. **OPEC Cumulative Surplus** (end of period)							k/
IBRD j/	62.4	131.5					403
Enders							337.0
W. J. Levy	75.0	122.0	191.0	264.0	337.0	402.0	449.0
Morgan Guaranty	80.0	137.0	191.0	231.0	248.0	235.0	179.0
Irving Trust							
- Prices rise	85.0	149.0	200.0	234.0	258.0	265.0	248.0
- Prices break	85.0	149.0	200.0	208.0	187.0	128.0	22.0
Citibank j/	66.0	102.0	139.0	169.0	188.0	196.0	189.0
CIA	83.1						

Note: Citibank figures are for its "central scenario."

a/ Morgan Guaranty export volumes and per-barrel revenues for 1975-1979 not available.
b/ Irving Trust assumes same export volumes in their Case A (rising prices) and Case B (price break in 1977).
c/ Data for Irving Trust and Citibank are prices rather than per-barrel government revenues.
d/ Actual cash receipts. Assumes a payments lag of $10 billion in 1974 and a further $4 billion in 1975, with a payments lag of $14 billion remaining constant thereafter.
e/ Data for Irving Trust and Citibank are total value of oil exports rather than total government oil revenues.
f/ Data for non-oil exports and investment income are not shown separately in the Citibank scenario.
g/ For Irving Trust, imports are net merchandise imports plus net services excluding interest. Assumed to be the same for both Cases A and B.
h/ Citibank imports of goods and services include foreign oil-company profits.
i/ Citibank data are "transfers."
j/ Excludes pre-1974 accumulation.
k/ IBRD notes that minor changes in assumptions could change 1980 cumulative surplus from $400 billion to $200 billion.

Sources:
1) Morgan Guaranty -- World Financial Markets, January 21, 1975.
2) Irving Trust -- The Economic View from One Wall Street, March 20, 1975.
3) First National City Bank (Citibank) -- Monthly Economic Letter, June 1975.
4) W. J. Levy Consultants Corp. -- Future OPEC Accumulation of Oil Money: A New Look at a Critical Problem, June 1975. This report summarizes the estimates of the above three reports.
5) Enders --"OPEC and the Industrial Countries: The Next Ten Years," Foreign Affairs, July 1975.
6) CIA -- Project World Oil Demand and OPEC Current Accounts, 1975-77, March 1975, updated informally, June 1975.
7) IBRD -- Capital Requirements of Developing Countries, April 28, 1975, as revised informally in July 1975.

Reproduced with permission of the author: Alfred Reifman, "U.S. Energy Policy: A Perspective on Major Issues" (Washington, D.C.: Congressional Research Service, U.S. Library of Congress, July 24, 1975).

Exhibit 12: Publicized Eurocurrency Credits: 1974 and First Half of 1975
($Millions and Percent to IBRD Members)

	Total	1974[1] I	1974[1] II	1974[1] III	1974[1] IV	1975 I	1975 II
TOTAL: IBRD MEMBERS	$26,520 100.0%	$5,975 100.0%	$12,078 100.0%	$3,891 100.0%	$3,332 100.0%	$2,292 100.0%	$4,165 100.0%
DEVELOPED COUNTRIES[2]	16,915 63.8	3,406 57.0	8,959 74.2	2,423 62.3	1,668 50.1	1,090 47.6	825 19.8
LESS-DEVELOPED COUNTRIES	9,605 36.2	2,569 43.0	3,119 25.8	1,468 37.7	1,664 49.9	1,202 52.4	3,340 80.2
OPEC Countries[3]	773 2.9	322 5.4	166 1.4	62 1.6	48 1.4	134 5.8	1,238 29.7
MSA Countries[4] ("Fourth World")	588 2.2	313 5.2	95 0.8	133 3.4	50 1.5	8 0.3	101 2.4
Cameroon	10	—	—	10	—	—	—
Egypt	230	50	80	100	—	—	30
El Salvador	50	—	15	—	35	—	24
Guyana	15	—	—	—	15	—	35
Ivory Coast	63	60	—	3	—	—	—
Pakistan	—	—	—	—	—	—	—
Sudan	220	200	—	20	—	8	12
Other Non-Oil LDCs[5] ("Third World")	8,244 31.1	1,934 32.4	2,858 23.6	1,273 32.7	1,566 47.0	1,060 46.2	2,001 48.0
(OTHER BORROWERS	2,103	475	316	684	187	594	764)

1 Credits for which the date of completion is not available are excluded from quarterly data but included in 1974 total. These amounted to $1,686 million for all borrowers.
2 Of which U.K. $5,897; France $3,304; Italy $2,240; and U.S. $1,439 million in 1974.
3 Of which Indonesia $349 million in 1974, and $992 in 1975 I & II; Venezuela $200 and Algeria $100 in 1975 II.
4 Noted here are all borrowing during this period from MSAs among the 41 listed in Exhibit 2.
5 Of which Brazil $1,668 in 1974 and $705 in 1975 I & II; Mexico $1,478 in 1974 and $619 in 1975 I & II; and Spain $1,169 in 1974.

Source: World Bank, "Borrowing in International Capital Markets," EC-181/752, August 1975, pp. 46–48.

Appendix Chart I

Main OECD Countries' Exports to OPEC, 1970-74

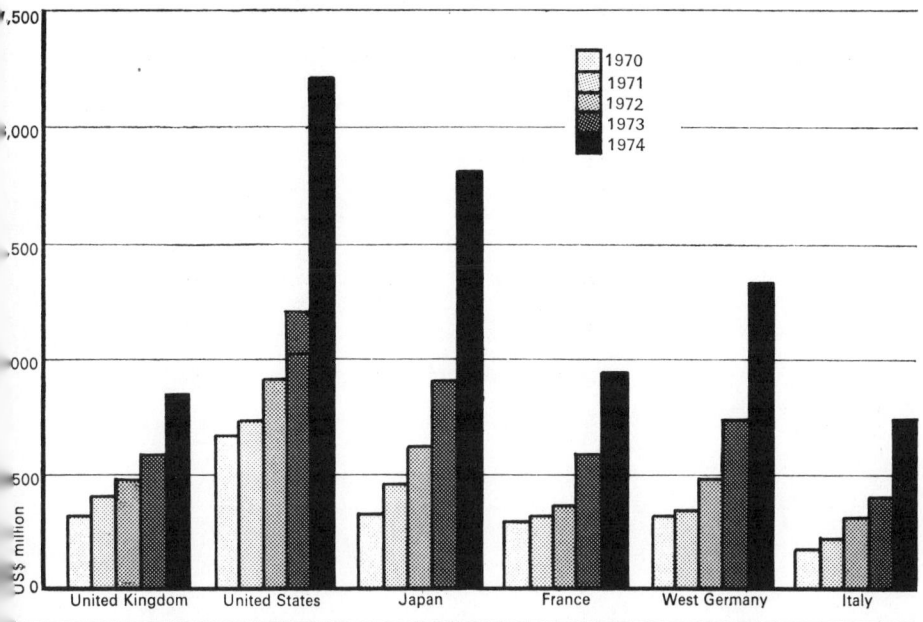

Source: *Trade & Industry*, July 11, 1975, Departments of Trade and Industry, London.

Appendix Table I

Value of Exports to OPEC Countries from Selected OECD Countries, 1973 & 1974

OECD Countries	$ Billion		% Increase
	1973	1974	
United Kingdóm	$ 1.75	$ 2.55	45.7%
United States	3.60	6.69	85.8
Japan	2.72	5.44	100.0
France	1.73	2.83	63.5
West Germany	2.26	4.04	78.7
Canada	0.26	0.58	118.9
Italy	1.21	2.23	84.4
Subtotals	$13.53	$24.36	80.0

Source: OECD Foreign Trade Statistics Series A and *The IMF Directory of Trade*.

Appendix Table II

Some Middle Eastern OPEC Country Defence Expenditure Contracts Reported in 1974 and Early 1975

Country	Weapon	Country	$m
Abu Dhabi (UAE)	Rapier Surface-to-Air AD Radar	UK	81
Iran	400 Chieftan tanks	UK	(500)
	6 963 class destroyers	US	660
	36 F-4E fighters	US	150
	1 Communications system	US	500
	6 Boeing 747 tankers	US	130
	80 Tomcat/Phoenix (F.14) fighters	US	1850
Iraq	60 Alouette helicopters	Fr	na
	40 Mig 23 fighters	USSR	268
Kuwait	36 Skyhawks	US	250
	18 Jaguar fighters	UK	63
Libya	T 6 Medium tanks, artillery and AD equipment	USSR	2000
	36 Mirage FI fighters	Fr	na
Qatar	20 Cascavel armoured cars	Brazil	na
Saudi Arabia	12 Jaguar fighters	UK	104
	Rapier SAM	UK	108
	200 Tanks		
	200 Armoured cars		
	AA guns and SAM system	Fr	825
	AA guns	Italy	196
	Tanks, guns, SAM	US	335
	Naval equipment and training	US	1500
	60 F-5E fighters	US	750
	20 Destroyer escort patrol boats	US	823

Sources: *The Defense Monitor*, Center for Defense Information, Washington, D.C., May 1975, and press reports.

Recipients of OPEC Bilateral Aid, 1974	Commitments ($7,172 mil.)	Disbursements ($2,031 mil.)
Islamic Nations	82%	90%
Arab nations	68	71
"Front Line": Egypt & Syria	58	54
Other Arab	10	17
Other Major Islamic: Pakistan & Bangladesh	14	20
Non-Islamic Nations	18%	10%
India	13	4
Others	5	6

Source: Exhibit 7B.

• *Short-term Swap Lines*, available to central banks for many years, were renewed in 1974.

• The *Community Loan Fund* of $3 billion was authorized early in 1975 by the European Community to recycle petrodollars borrowed for this purpose from OPEC countries. These would be made available on normal terms to Community members with oil-induced payments problems and who found themselves unable to borrow sufficient amounts through regular channels. In mid-1975, the first applications to this Fund were being made by Ireland and Italy.

• The major innovation in this category is the $25 billion *OECD Financial Support Fund*, which was conceived and developed under a variety of names.[7] This Fund, awaiting legislative approval by OECD countries, is intended to be the last of the last resorts. An applicant country seeking loans must demonstrate that it is encountering serious financial difficulties; that it has exhausted other resources such as its own reserves or loans on reasonable terms (including those offered by the IMF and the Community Loan Fund); that it is taking measures to conserve energy and develop alternative energy sources; and that its

---

7 The OECD Financial Support Fund was originally called the "Kissinger Plan" or the "Kissinger-Simon Plan," though it deserves to be associated as well with Emile Van Lennep, the Director of the OECD. More officially, it became known first as the "Financial Solidarity Fund," but finally the "Financial Support Fund" to avoid implying a confrontationist character. The phrase, "safety net" has also been used but soon became ambiguous, other "safety nets" having meanwhile been placed beneath other actors in the recycling operation.

economic policies include measures to redress its external financial situation. These standards being met, loans will be granted from a common pool, the risks being shared among all participants in proportion to their quotas.

Other Beneficial Policies and Developments

Besides creating these new recycling mechanisms, both the importing and OPEC countries have adopted other policies which, helped by some unplanned events, have acted to ease one or more of those financing problems that were perceived during much of 1974.

... in the Importing Countries

From the outset of the crisis, the major importing countries strove to quench any tendency among them to seek unilateral remedies that promised to be harmful to all. In January 1974, the 20 governments then negotiating monetary reform pledged to avoid trade restrictions. The Washington Energy Conference in February led to the creation of the 18-member International Energy Agency, with specific programs to deal with emergency situations. Meanwhile, the IMF made restraint from unilateral restrictive measures a condition for access to the IMF Oil Facility.

The importing countries have also taken steps to curtail demand for oil.

The major factor behind reduced OPEC oil production, however, has been an unplanned development—widespread recession among the importing countries.[8]

Also unplanned, but more truly fortuitous, have been two successive mild winters, easing demand for heating oil, an especially large end-use for European petroleum imports.

... in OPEC Countries

While these developments in the importing countries have tended to diminish OPEC oil revenues somewhat, greater than

---

[8] Daily OPEC production was down to about 26 million barrels per day during the first half of 1975, as against 30 million in 1974 and almost 33 million in September 1973, just before the oil crisis.

Initial Problems

anticipated absorption of these earnings is reducing their petrodollar surpluses. OPEC's own imports—"direct absorption"—rose by 90 percent in value and by 40 to 45 percent in volume between 1973 and 1974.[9] "Indirect absorption" has also grown sharply.

As for the deployment of now-reduced petrodollar surpluses, OPEC countries have undertaken a considerable volume of direct recycling, either to the IMF Oil Facilities or via bilateral aid. Meanwhile, OPEC's placement of its surpluses among intermediaries appears to be shifting from short-term deposits on call to longer-term investments. This reflects in part a lowering of short-term rates, but also growing knowledge and confidence by OPEC countries in making longer-term investments.

---

9 World Bank, "Capital Requirements of Developing Countries," April 1975, p. 19, as quoted in Alfred Reifman, "U.S. Energy Policy: A Perspective on Major Issues" (Washington, D.C.: Congressional Research Service, U.S. Library of Congress, July 24, 1975), p. 13.

III. Perceptions to 1980

The developments just mentioned have eased the initial anxiety over the four challenges first seen. But, what about the worldwide financial implications of the 1973-74 price rise over the rest of the decade? While present perceptions have not settled—indeed, they often disagree—it is useful to review them in relation to three issues.

Contradictory Signals for OPEC Prices

One of these, of course, is what happens to the OPEC price between now and 1980. Estimates diverge sharply, as seen by the over 2:1 discrepancy among those given for 1980 in Exhibit 11-B.[1] Two of the five accept the possibility that OPEC prices will fall after 1977 in current prices, meaning a yet sharper decline in real terms.

Such diverse forecasts stem from widely differing assumptions regarding several prospective developments that could affect world demand for OPEC oil one way or the other (note the variety of estimates in 11-A), or that by other means could influence OPEC pricing policies later in this decade.

Favoring lower prices before 1980 we see that ...

●A number of noncommunist importing countries appear likely to provide more of their own oil before 1980, one estimate seeing a 40 percent increase.[2] Britain, Norway and Mexico are expected to reach self-sufficiency by then and even begin exporting.

●The same estimate foresees China supplying Japan with some of its needs by 1980 and there is talk of the Soviet Union exporting oil to the United States.

---

1 The Citibank entries throughout Exhibit 11 are for its "Central Scenario." Not given there are three companion scenarios, one of which assumes a sharp break in OPEC prices after 1977.

2 This estimate (in Reifman, "U.S. Energy Policy," p. 10) foresees oil production from non-OPEC sources which averaged 15 million barrels per day in 1974 and 1975, rising by some 6 million barrels per day in noncommunist countries, with an additional one million barrels to be exported from China to Japan. After 1980, further alternatives to OPEC oil appear possible, a major example being U.S. production from its Outer Continental Shelf and its Naval Petroleum Reserve #4 in Alaska.

Present Problems 19

●Especially in the United States, where the traditional low cost of energy has encouraged its wasteful use, there remains considerable potential for further curtailing oil consumption. In the light of progress to date, however, the prospect for effective U.S. action is not particularly bright—though it might be stimulated by the recent 10 percent rise in OPEC oil prices.

●Any prolonged or significant decline in demand for OPEC oil might severely test the key to maintaining its price, the will of its diverse members to implement production-sharing arrangements under such circumstances.

Favoring higher prices, we see that . . .

●Mounting absorption of their earnings may force some OPEC countries into annual current account deficits sooner than they expected, generating pressure within them for yet further price increases beyond that of September 1975.

●Recovery from the present degree of worldwide recession is bound to stimulate higher oil imports. Moreover, recovery might also revive inflation accompanied by export price increases that could encourage emulation by OPEC countries.

●There is no discernable reason why forthcoming winters should be as unusually mild in the major consuming countries as were the last two.

●The dominant OPEC country from Group I, Saudi Arabia, has demonstrated its willingness to curtail oil production significantly, suggesting reduced chances that even were demand for OPEC oil reduced, this could be done sufficiently to force down the OPEC price.

A New Phase in the Problems Facing Industrialized Countries

. . . from handling OPEC surpluses . . .

As we have seen, the July 1974 World Bank projection of OPEC surpluses (accumulating $650 billion in current dollars by 1980 and increasing annually by over $100 billion to exceed $1,200 billion in 1985) heightened the initial concern over financing higher oil prices. More recent projections, developed throughout Exhibit 11 and culminating in 11-H and 11-I, are considerably lower, especially those of three New York banks which show the OPEC aggregate surplus peaking between 1978 and 1980. This "optimistic" view is contested, however, in the Levy projections.[3]

While these recent estimates disagree, it is safe to say that they have eased to a greater or lesser extent the initial concern over the dimensions of the financing problems. For, especially when these figures are deflated into real terms by assumptions of inflation, they foresee the importing countries being released sooner or later from relentless growing indebtedness to OPEC.

... to handling increasing transfers of real resources

There is another side of the coin, however. As their individual surplus accumulations peak in different years,[4] OPEC countries will begin disinvesting in the importing countries. They will cash in for imported goods and services not merely their current earnings from oil sales and repatriated investment earnings but also some of their stocks of accumulated credits (11-I). For the industrialized oil-importing countries—the only ones in a position to furnish OPEC with its desired goods and services—this represents a shift from the present phase of their transferring financial claims to OPEC countries to one marked by ever larger transfers of real resources.

This new phase presents both problems and opportunities to the industrialized countries, the degree of each depending largely on the timing and extent of the shift, especially if it is sudden and

---

3 The study by W.J. Levy and Consultants, cited under Exhibit 11—and, in fact, providing the basis for that table—presents an illuminating analysis of techniques and assumptions of other estimates, notably those of the three New York banks. Another thoughtful discussion is offered by J. Michael Jefferson in "Conflict and Cooperation in an Era of High Oil Prices," a paper delivered at the University College at Buckingham Conference in June 1975. The Conference papers, as edited by Professor Berstein, will be published in Britain by MacMillan in 1976. Both Levy and Jefferson give considerable weight to the increasing volume of remitted earnings from OPEC investments as a supplement to their oil revenues (Exhibit 11-F).

4 The Citibank's "Central Scenario," whose forecasts for all OPEC countries are included in Exhibit 11, predicts that the accumulated surpluses of individual OPEC countries will peak as follows (values in $billions): *1973*: Algeria $-2.6; *1974*: Libya $4.0; *1975*: Indonesia $-3.0; *1976*: Iran $13.0; *1977*: Iraq $11.3 and Ecuador $0.6; *1978*: Venezuela $24.7 and Abu Dhabi $4.6; *1980*: Saudi Arabia $90.4, Kuwait $47.1 and Nigeria $25.4; and *1985*: Qatar $14.9. As Exhibit 11-I shows, this forecast has the accumulated surpluses of all 12 countries peaking in *1979* at $196.2 billion.

sharp.[5] The crucial factor is how it interacts with their then-current states of inflation and recession and the readiness of certain industrial sectors to furnish what OPEC wants at the right price and delivery schedules.

Some concern has been expressed on the latter point. In the Statement accompanying this report, the BNAC urges that where possible, surplus petrodollars be recycled into productive investment rather than into consumption. Private proposals for how this might be done have appeared from time to time.[6] The most publicized is a *Foreign Affairs* article of international authorship that suggests OPEC Mutual Investment Trusts to channel surplus petrodollars into productive investment in the oil-importing countries in ways mutually beneficial and protective to them and to the oil producers alike.

The Situation Facing the LDCs

The financing problems of the LDCs over the remainder of this decade have become a major theme for eloquence and action. Without attempting to cover this expanding topic, let us note merely some present thinking about causes, dimensions and appropriate responses.

For one thing, the range of economic difficulties facing the LDCs are being seen less exclusively linked to the oil price shock of 1973-74. Causal factors are now taken to include higher import prices of food, fertilizer and manufactured goods. In addition, the deep and prolonged recession within the developed countries has depressed export growth and earnings among those more fortunate LDCs with something to sell. Correspondingly, the concept of recycling to LDCs is not seen merely in terms of surplus petrodollars, as indicated in Exhibit 10 by greater commitments by non-OPEC countries with strong payments positions to funding

---

5 The implications of how rapidly the transfer of real resources takes place are discussed in "The Absorptive Capacity of OPEC Countries," released by the U.S. Treasury Department, September 5, 1975.

6 An early proposal was by Elizabeth Monroe and Robert Mabro, "Arab Wealth from Oil: Problems of its Investment," *International Affairs*, Chatham House, London, January 1974. The *Foreign Affairs* article (January 1975) was "How Can the World Afford OPEC Oil?," by Khodadad Farmanfarmanian (Iran), Armin Gutowski (Germany), Saburo Okita (Japan), Robert Roosa (U.S.), and Carroll L. Wilson (U.S.). A more recent proposal for an "OPEC Endowment Fund Trusteeship" has been made by Robert M. MacIntosh, Executive Vice-President of the Bank of Nova Scotia.

the IMF's Second Oil Facility. Also, there are signs that this IMF operation might well be its last oriented specifically to meeting oil-induced deficits.

Opinions differ as to whether the OPEC price rise of 1973-74 should still be regarded as the most important root cause of difficulties facing the LDCs, but there is no disagreement as to their seriousness. Reviewing immediate prospects, the IMF's Annual Report for 1975 sees the following developments in the world's payment situation:[7]

	Current Account Balances ($ Billions)		
	1973	1974	1975
Major oil exporters	6	70	50
Industrial countries	10	-12	1
Non-oil primary producing countries			
More developed	1	-12	-12
Less developed (LDCs)	-9	-28	-35
Total	8	19	4

Looking further ahead, the World Bank initiated in 1975 a study of the capital requirements of the LDCs "... to maintain a reasonable rate of growth in per capita income for the remainder of the decade." This unpublished study concludes that substantial increases in capital flows, official and private, will be essential to sustain even minimum acceptable growth rates during 1976-80.

Similar concerns are being expressed by individual governments and private observers. A number of major policy proposals are emerging, notably those developed within the IMF and World Bank and that of the United States presented on September 1 before a special United Nations General Assembly. These various

---

7   Source: International Monetary Fund, *Annual Report* for the Fiscal Year ending April 30, 1975 (Washington, August 1975), p. 16. The 1975 projections are "subject to considerable uncertainty and should be viewed as rough orders of magnitude." *Industrial Countries* include the United States, Canada, Japan, the European Community countries minus Ireland, and Norway, Sweden and Switzerland. The *More Developed Non-Oil Producing Countries* comprise Australia, Finland, Greece, Iceland, Ireland, Malta, New Zealand, Portugal, South Africa, Spain, Turkey, and Yugoslavia. The *Less-Developed* subgroup comprises all other countries, and corresponds to the LDCs.

plans appear to serve one or both of the two main objectives: (1) increasing capital flows to LDCs through lending by governments and by multinational institutions, and through improving their access to private capital markets; and (2) increasing and stabilizing their export earnings.

The call for the first broad objective signals another important aspect of present thinking: that the accomplishments of the past two years in arranging to recycle surplus petrodollars to LDCs will not meet their needs over the remainder of the decade.

•For the MSAs the answer is seen to lie in expanding directed recycling on more concessional terms. In part, this would be done through various forms of bilateral aid from the developed countries and the OPEC countries—the "Old Rich and the New Rich." Here the World Bank, anticipating the gradual disappearance of OPEC surpluses over the period, sees the developed countries needing to contribute an increasing share. The other major vehicles for directed recycling are, of course, the World Bank and the IMF, borrowing for the purpose mainly from the developed and OPEC countries. These organizations are now planning to expand and supplement their facilities. emphasizing more concessional terms.

•The more favored "Third World" LDCs are expected to find borrowing increasingly a permanent means of adjustment. But, since they have relied heavily on natural recycling, the question arises for how long this will hold up. While the $2 billion in Eurocurrencies borrowed from within this group in the second quarter of 1975 (Exhibit 12) is encouraging, there is increasing concern that more and more of them will find that borrowing on international credit markets is "an option available to a limited group of countries in 1974 and to even fewer in 1975 because of less favorable export positions and strained external debt-carrying capacities."[8] Already, banks report more urgency among borrowers and more reluctance among lenders. It is mainly for these countries that proposals are coming forth to improve their access to private capital markets.

---

8   *IMF Annual Report*, p. 22.

APPENDIX

The Disposition of OPEC Oil Revenues[1]

by Susan Hart

This Appendix sets out how OPEC countries are using their greatly increased oil revenues. As noted in the main paper, there are two basic uses:
Absorption: investment by OPEC countries in their own internal development, principally through the purchase of goods and services from oil-importing countries, and
Primary Recycling: investment of the remaining surplus petrodollars to the oil-importing countries—both developing and developed—via a variety of financial transfer operations.

Absorption of Oil Revenues by Buying Foreign Goods and Services

The capacity of individual OPEC countries to absorb their growing oil receipts (estimated at over $90 billion in 1974) by importing goods and services differs significantly among them. In the long run, any surplus petrodollars will be concentrated in a group of high-income, low absorption countries such as Saudi Arabia, Kuwait, Abu Dhabi, the United Emirates, and Qatar (Group I in Exhibit 3 of the main paper). Countries in Group II such as Venezuela and Iran may return to situations of persistent current account deficit well before the end of the decade; Algeria is reported to be borrowing to balance her external account. Group III countries (Nigeria and Indonesia) too are likely to spend all they can.

In the period 1973-74 of rapidly increasing oil revenues, the low as well as the high absorbers have made significant efforts to increase the level of their imports and to step up internal investment programmes. Had it not been for the fact that the *value* of aggregate OPEC imports in 1974 increased by as much as 50 percent over the previous year, many oil-importing countries would have faced accelerating trade deficits, and world banking institutions would have had to channel even greater amounts of surplus OPEC wealth. Chart I indicates the increase in exports to OPEC countries from six industrialised countries since 1970, and Table I illustrates the growth of such exports from seven industrialised countries between 1973 and 1974.

Military Purchases

The *real* increase in imports as opposed to the value of OPEC countries in 1974 is estimated to have been 40 percent over 1973. Little reliable data exists on the categories of imports for each country, but according to *The Defense Monitor* a significant proportion of the estimated $32.5 billion has been spent on military hardware and systems.[2] This same source suggests that

---

[1] Compiled separately from the main paper in the BNAC's London office under the direction of Simon Webley.

[2] *The Defense Monitor*, Center for Defense Information, Washington, D.C., May 1975.

the United States alone sold arms worth $4.4 billion to Persian Gulf countries with the United Kingdom, Russia and France supplying military equipment worth a further $3.5 billion. The major buyers were Saudi Arabia, Iran, Kuwait, Iraq, and Oman. Table II lists some reported defence contracts in 1974 and the early months of 1975. These were clearly not all paid for at the time of ordering, but they give some indication of the size and value of this type of import. It should also be noted that some defence expenditures by OPEC countries are on behalf of so-called "front-line" Middle Eastern countries (Egypt, Syria and Jordan) and this indirect absorption is sometimes recorded as aid.

Development Plans

Most OPEC countries have ambitious programmes for longer-term commercial and social development. A number of them have drawn up five-year plans which they are implementing swiftly. Iran's current programme 1973-78 has been increased by $36 billion to $69 billion and Nigeria's 1975-80 plan envisages spending $53 billion—an increase of 200 percent over the figures in the first draft of the plan. The Saudi Arabians are completing their current programme and have announced that in 1976-80 they plan to spend $142 billion on internal development.

The plans of OPEC countries contain, and to some extent rely on, joint ventures with companies in the industrialised countries. For instance, a leading West German aerospace and transport concern is establishing a joint enterprise with Iran to develop regional transport systems there, while a U.S. company has undertaken to build a natural gas treatment plant in Algeria.

According to a Kuwaiti spokesman, OPEC countries will give top priority to the long-term problem of food production. In this context, energy and food related industries will receive primary attention. Investment in petrochemical and fertiliser enterprises has already begun. In July 1974, 10 Middle Eastern nations formed the Arab Capital Investment Corporation specifically for developing alternative energy sources both within and outside the Middle East area.

One commentator has suggested that by 1985, Arab ethelyne capacity would contribute an increase of up to 10 percent in world ethelyne production.[3] This and other petrochemical projects will require investment of perhaps $15 billion to cover production forecasts of base chemical, methanol and plastic material alone. When the planned oil refinery projects and gas production plants are included, the estimated capital investment figure climbs to $40 billion.

While available information illustrates an impressive picture of OPEC improvement activity in the recent past, several factors suggest that a cautious view should be taken of long-term prospects. First, population both in absolute size and in levels of literacy affects capacity to absorb. Often the countries with the largest oil reserves have the smallest populations. Second, the amount of petroleum reserves affects the level of present and future oil

---

[3] Ray Dafter, 'The Oil Producers Move On Downstream,' *The Financial Times*, November 1974.

Appendix 27

production and consequently of anticipated revenues. Saudi Arabia, with the largest known reserves, small population, and unsophisticated social and economic structures, will long be a surplus revenue country; it simply will be unable to absorb its vast revenues rapidly. On the other hand, Indonesia is already finding its revenues and expenditures hard to balance and the Iranian Government anticipates that it will be running a current account deficit by the Autumn of 1976. Algeria is reported to be raising up to $1 billion to finance its 1975 balance of payments and Iraq is known to be trying to raise a loan on the international finance market. Third, the volume of imports reaching intended destinations within OPEC countries is limited by inadequate port facilities and ancillary transport networks. Such bottlenecks can be overcome as ports are expanded, roads constructed and transportation vehicles purchased or produced, but all of these activities take time under any circumstances.

There is some basis for suggesting, therefore, that the circumstances of 1974 were particularly favourable for a sharp increase in exports of goods and services from oil-consuming to oil-producing countries. The unexpectedly high level of OPEC imports contributed to a lower than expected level of surplus OPEC revenues for the year. However, it would be imprudent to project future OPEC imports and future OPEC revenues on the basis of 1974 rates.[4]

Primary Recycling

The substantial quantity of petrodollars remaining after OPEC's purchases of goods and services enter the primary recycling phase as initial placement. Exhibit 6 in the main text indicates the distribution of the $56.2 billion surplus funds in 1974.

Approximately $32 billion (57 percent) of the surplus was deposited in the United States or the United Kingdom. A further $9 billion (16 percent) was deposited in other countries in currencies other than dollars or sterling, and a proportion of the remainder ($7.5 billion) was utilized as direct aid or loans to developing countries—some of which went via the international organisations.[5]

Primary Recycling to Developing Countries

The scope and distribution of assistance by OPEC countries to developing countries is contained in figures published by the Development Assistance Committee of the OECD.[6] In 1974, dispersements of OPEC

---

[4] For some detailed estimates of absorbative capacity of OPEC countries 1974-77, see 'Conflict and Cooperation in an Era of High Oil Prices,' a lecture delivered by J.M. Jefferson at the University of Buckingham Conference, London, June 1975.

[5] Some preliminary estimates in Exhibit 6 for 1975 show a shift in deposits from the United Kingdom and the United States to other countries.

[6] *OECD Observer*, March/April 1975.

official development assistance totalled $2.54 billion or 1.8 percent of their aggregate GNP. This compares with $11.3 billion or 0.33 percent of GNP of 17 OECD countries. If portfolio and direct investments are included, then OPEC aid represents 3.4 percent of GNP (4.75 billion) compared with 0.77 percent of OECD GNP ($26.4 billion).

It is apparent from these figures, together with pronouncements from various OPEC leaders, that there is a continuous interest in directing a proportion of OPEC financial resources to developing countries. Initially, this attention has been directed to other Arab or Islamic nations, principally Egypt, Syria and Pakistan, though India received a considerable sum. These four countries accounted for 78 percent of 1974 flows.

Discussion of long-term plans for redeployment and mobilisation of Arab money for use in the Arab and to some extent the Islamic world dates back at least to the March 1973 meeting of the Council of Arab Economic Unity. Subsequently, it was decided that a study should be made to determine the best ways to implement this programme.

One billion dollars was given to Egypt and other Arab 'front line' states in 1973 and 1974, primarily for defence and reconstruction purposes. In May 1974, a charter was completed for a 24-member Islamic Development Bank which will extend loans on an interest-free basis. Commitments to existing regional development institutions such as the Kuwait Fund for Arab Economic Development, the Arab Fund for Economic and Social Development and the Islamic Bank have been increased.

However, the OPEC countries with surplus oil revenues have not restricted their aid only to Arab and Islamic countries. The total external debts of the UN-designated less-developed nations totalled nearly $100 billion at the end of 1972 and these debts have been rising rapidly. Oil import costs for these countries reached $14.6 billion in 1974 compared with $5.9 billion in 1973. Debts and debt servicing now account for nearly 17 percent of their worldwide non-oil export earnings.

Immediate steps have been taken by OPEC countries to assist the most seriously affected countries (MSAs). About one-third of bilateral aid went to them. OPEC commitments to the first IMF special oil facility which was available for use by MSAs totalled SDR 2.65 billion ($3.3 billion) in 1974 (see Exhibit 10-A in the main paper), and in addition they have provided over $2 billion to the World Bank. Some individual oil-producing countries have made other arrangements; Venezuela, for instance, is negotiating concessional lending arrangements with the Inter-American Bank and the Caribbean Development Bank. Iran has concluded bilateral arrangements involving soft loans for project aid and financing arrangements of some $1.5 billion to be disbursed over the next three-five years. Exhibit 10-B in the main text indicates which countries received the bulk of OPEC aid in 1974.

At the Rome World Food Conference in 1974, U.S. Secretary of State Henry Kissinger called upon the oil-wealthy states to assume a far greater responsibility for feeding the world's poor. OPEC countries responded by drawing up general plans for a major international agricultural fund to channel investment into developing countries, but as yet no figure has been suggested. OPEC concern for the less-developed countries is expressed both in its insistence that any oil producer/oil consumer conference include discussion of all raw materials as a matter of priority and in an eloquent Solemn

Declaration which emanated from a Conference of the Sovereigns and Heads of State of the OPEC member countries in March 1975.

Primary Recycling to Developed Countries

It is here that the greatest amount of primary recycling of petrodollars has taken place.

... through Capital Markets

About 35-40 percent of OPEC surplus petrodollars entered the *Eurocurrency market* primarily on a short-term basis. As short-term rates began to fall toward the end of 1974, OPEC investors sought longer-term assets. Arab participation as managers and underwriters in bond market activity began to increase at that time. In 1974, *Eurobond* issues made in Kuwaiti dinars totalled perhaps as much as $51.3 million in dollar equivalents, indicating growing acceptance of Middle Eastern currencies as a medium for raising funds.[7] Furthermore, Arab institutions have shown a willingness to assist Third World countries to raise fixed rate capital in international markets, as indicated by a two-tranche $25 million issue raised for Brazil but arranged solely by Kuwaiti institutions.

U.S. and U.K. capital markets provide a major source of liquid instruments of public debt, a preferred outlet for the funds of Arab central banks. A portion of OPEC governments' surpluses are held in highly liquid and secure form to make up part of their official international reserves. The increase in assets by OPEC countries recorded as official reserves was about $35 billion in 1974, held primarily in dollars.

The $6 billion placed by OPEC in U.K. sterling assets and government securities in 1974 (see Exhibit 6, lines 1-4) offset almost exactly the increase in the value of Britain's net oil imports. The $11 billion which flowed directly into U.S. banks and purchases of government securities (Exhibit 6, lines 7-9) financed roughly two-thirds of the increase in the U.S. current account deficit with OPEC and was considerably in excess of the $4 billion overall current account deficit for the United States in 1974.

All in all, capital markets expanded substantially as a result of OPEC capital inflows in 1974. By the end of March 1975, oil-exporting countries' deposits accounted for 12.9 percent of all foreign currency deposits in London, compared with only 4.4 percent in 1973. Exchange reserves held in sterling by oil-exporting countries accounted for 68 percent of sterling exchange reserves held by central monetary institutions in March 1975.

The purchase of government bonds and securities is a comparatively safe avenue for OPEC investment. Of the $6 billion in direct claims on the United Kingdom by OPEC, $900 million was in British Government stocks and $2.7 billion in U.S. Treasury bills (see Exhibit 6).

Six billion dollars of the $11 billion which flowed into the United States was used to purchase government and agency securities. Saudi Arabia, for instance, was a substantial purchaser of special nonmarketable government securities.

---

7 Michael Blanden, 'London's Role Sustained,' *The Financial Times*, March 3, 1975.

Iran chose a different method of recycling its surplus revenues to developed countries. It extended over $5 billion in bilateral loans to three major European countries in 1974, not all of which was actually drawn: Britain, $1.2 billion; France, $1 billion; and Italy, $3 billion. These were negotiated at current rates.

... through Direct Investment

OPEC investors have so far been reluctant to place much of their earnings in foreign direct investment. However, in times of rapid inflation, direct investment can provide some hedge against depreciation of monetary assets.

Middle Eastern OPEC countries have made some notable purchases in the past two years and the Kuwaities, who have the most investment experience, have been the most active. There have even been a few cases of Arab purchases of controlling interest in small U.S. banks, though one attempt to purchase a Detroit bank was thwarted.

Real estate purchases have attracted considerable publicity. Kuwait has acquisitions in California, Massachusetts and Virginia. Iran and Abu Dhabi have made some purchases too. Kuwait will develop a residential resort on Kiawah Island off the South Carolina coast and the Sheikh of Abu Dhabi bought a £36 million stake in London's Commercial Union office building.

The publicity that these deals have received and the quick reaction to them tends to exaggerate the importance of direct investment by oil-producing nations. It still accounts for a very small proportion of OPEC surplus revenue disposition and probably does not exceed a few billion dollars in value. The vociferousness of some sections of the U.S. public on this matter was echoed in Congress by a series of restrictive Bills in 1974 and 1975. However, the Administration has so far successfully resisted any action on the matter.

... through Equity and other Portfolio Investment

In 1975, it became clear that Middle Eastern OPEC countries were interested in financing fixed interest debts of first class companies, other than oil companies. The American Telephone and Telegraph Company placement of $100 million six-year notes at 8¼ percent with the Saudi Arabian Government is an illustration of this interest.

Investment in major western industry equity has caused some controversy. In 1974, Kuwait bought for £107 million a controlling interest in St. Martin's Property Company of the United Kingdom, and Iran purchased a 25 percent interest in Germany's Krupp Steel. When Kuwait bought a 14 percent share of Daimler-Benz, the Deutsche Bank responded by buying 29 percent of the shares, bringing its total holdings to well over 50 percent, and the German Government began to talk of imposing restrictions on foreign investors.

The Bank of England estimates that OPEC investment in holdings of U.K. equities and property does not exceed $700 million and similar investment in the United States does not exceed $1 billion. Nevertheless, it is not very easy to distinguish the buyers of stock on the stock exchanges of the industrialised countries. Certainly the recovery in stock prices in 1975 is not easily explained without reference to substantial demand for stock outside indigenous sources.

As the 1975 Economic Report of the President of the United States points out, the relative attractiveness of the U.S. economy makes it highly likely that the country will be the recipient of substantial amounts of long-term investment funds in the long run.

Members of the British-North American Committee

Chairmen
SIR DAVID BARRAN
A Managing Director, Shell Transport and Trading Company Ltd., London

NICHOLAS J. CAMPBELL, JR.
Director and Senior Vice President, Exxon Corporation, New York, New York

Vice Chairmen
IAN MacGREGOR
Chairman, AMAX Incorporated, Greenwich, Connecticut

RICHARD DOBSON
Chairman, British-American Tobacco Co., Ltd., London

Chairman, Executive Committee
W.O. TWAITS
Director and Vice President, Royal Bank of Canada, Toronto, Ontario

Members
*A. ROBERT ABBOUD
Deputy Chairman of the Board, The First National Bank of Chicago, Chicago, Illinois

W.S. ANDERSON
Chairman and President, NCR Corporation, Dayton, Ohio

J.A. ARMSTRONG
Chairman and Chief Executive Officer, Imperial Oil Limited, Toronto, Ontario

A.E. BALLOCH
Executive Vice President, Bowater Incorporated, Old Greenwich, Connecticut

DAVID BASNETT
General Secretary, General and Municipal Workers' Union, Esher, Surrey

ROBERT BELGRAVE
Policy Planning Advisor, British Petroleum Limited, London

RUSSELL BELL
Director of Research, Canadian Labour Congress, Ottawa, Ontario

C. FRED BERGSTEN
Senior Fellow, The Brookings Institution, Washington, D.C.

I.H. STUART BLACK
Chairman, General Accident Fire and Life Assurance Corporation Ltd., Perth, Scotland

HOWARD BLAUVELT
Chairman and Chief Executive Officer, Continental Oil Company, Stamford, Connecticut

HARRY BRIDGES
President and Chief Executive Officer, Shell Oil Company, Houston, Texas

DR. CHARLES CARTER
Vice-Chancellor, University of Lancaster, Lancaster

SILAS S. CATHCART
Chairman and Chief Executive Officer, Illinois Tool Works Inc., Chicago, Illinois

SIR FREDERICK CATHERWOOD
Director, John Laing & Son Ltd., London

HAROLD van B. CLEVELAND
Vice President, First National City Bank, New York, New York

KIT COPE
Overseas Director, Confederation of British Industry, London

WILLIAM DODGE
Ottawa, Ontario

ALASTAIR F. DOWN
Chairman and Chief Executive, Burmah Oil Company, Swindon

G. EASTWOOD
General Secretary, Association of Patternmakers and Allied Craftsmen, London

Committee Members

LORD FEATHER
London

JOSEPH B. FLAVIN
Executive Vice President, Xerox Corporation, Stamford, Connecticut

ROBERT M. FOWLER
President, C.D. Howe Research Institute, Montreal, Quebec

WILLIAM FRASER
Chairman, British Insulated Callender's Cables Ltd., London

DOUGLAS R. FULLER
Vice-Chairman, The Northern Trust Company, Chicago, Illinois

LORD GOODMAN
Messrs. Goodman, Derrick & Co., London

GEORGE GOYDER
British Secretary, British-North American Committee, London

EDWARD GRUBB
Chairman and Chief Officer, The International Nickel Company of Canada, Ltd., Toronto, Ontario

*HON. HENRY HANKEY
Director of Lloyds Bank, International Limited, London

AUGUSTIN S. HART
Group Vice President, Quaker Oats Company, Chicago, Illinois

HENRY J. HEINZ II
Chairman of the Board, H.J. Heinz Company, Pittsburgh, Pennsylvania

*JACK HENDLEY
General Manager (International), Midland Bank Limited, London

HARRY G. JOHNSON
Professor of Economics, University of Chicago, Chicago, Illinois

JOSEPH D. KEENAN
International Secretary, International Brotherhood of Electrical Workers, AFL-CIO, Washington, D.C.

TOM KILLEFER
Vice-President, Finance and General Counsel, Chrysler Corporation, Detroit, Michigan

LANE KIRKLAND
Secretary-Treasurer, AFL-CIO, Washington, D.C.

CURTIS M. KLAERNER
Executive Vice President and Director, Mobil Oil Corporation, New York, New York

*HECTOR LAING
Chairman, United Biscuits Limited, Isleworth, Middlesex

H.U.A. LAMBERT
Vice-Chairman, Barclays Bank Ltd., London

HERBERT H. LANK
Director, Du Pont of Canada Ltd., Montreal, Quebec

JOHN LAWRENCE
Chairman of the Board, Dresser Industries, Inc., Dallas, Texas

FRANKLIN A. LINDSAY
Chairman of the Board, Itek Corporation, Lexington, Massachusetts

JAMES LONGMORE
Director of Lloyds Bank, International Limited, London

JAY LOVESTONE
International Affairs Consultant, AFL-CIO, Washington, D.C.

RAY W. MACDONALD
Chairman and Chief Executive Officer, Burroughs Corporation, Detroit, Michigan

AUGUSTINE R. MARUSI
Chairman and Chief Executive Officer, Borden Inc., New York, New York

*CARGILL MacMILLAN, JR.
Senior Vice President, Cargill Incorporated, Minneapolis, Minnesota

Committee Members

B.J. McGILL
Vice President and General Manager,
International, The Royal Bank of Canada,
Montreal, Quebec

WILLIAM C.Y. McGREGOR
International Vice-President, Brotherhood
of Railway, Airline and Steamship Clerks,
Montreal, Quebec

DONALD E. MEADS
Chairman and Chief Executive Officer,
Certain-Teed Products Corporation,
Valley Forge, Pennsylvania

SIR PETER MENZIES
Chairman, The Electricity Council,
London

DEREK F. MITCHELL
President, BP Canada Ltd.,
Montreal, Quebec

JOSEPH P. MONGE
Vice Chairman, International Paper
Company, New York, New York

D.R. MONTGOMERY
Secretary-Treasurer, Canadian
Labour Congress, Ottawa, Ontario

DR. MALCOLM MOOS
Santa Barbara, California

JOSEPH MORRIS
President, Canadian Labour Congress,
Ottawa, Ontario

CHARLES MUNRO
President, Canadian Federation of
Agriculture, Ottawa, Ontario

KENNETH D. NADEN
President, National Council of Farmer
Cooperatives, Washington, D.C.

JOSEPH NICKERSON
Chairman, The Nickerson Group of
Companies, Rothwell, Lincoln

WILLIAM S. OGDEN
Executive Vice President,
The Chase Manhattan Bank,
New York, New York

WILLIAM R. PEARCE
Vice President, Cargill Incorporated,
Minneapolis, Minnesota

SIR RICHARD POWELL
Director, Hill Samuel Group Limited,
London

J.G. PRENTICE
Chairman of the Board, Canadian
Forest Products Ltd., Vancouver,
British Columbia

BEN C. ROBERTS
Professor of Industrial Relations,
London School of Economics, London

HAROLD B. ROSE
Group Economic Adviser,
Barclays Bank Limited,
London

WILLIAM SALOMON
Managing Partner, Salomon Brothers,
New York, New York

A.C.I. SAMUEL
Director, British Agrochemicals
Association, London

DAVID SCOTT
Chairman and Chief Executive Officer,
Allis-Chalmers Corporation, Milwaukee,
Wisconsin

PETER SCOTT
Chairman, Provincial Insurance
Company Limited, Stramongate,
Kendal, Westmoreland

LORD SEEBOHM
Chairman, Finance for Industry,
London

THE EARL OF SELKIRK
President, Royal Central Asian Society,
London

GEORGE P. SHULTZ
President, Bechtel Corporation,
San Francisco, California

G.F. SMITH
General Secretary, Union of
Construction, Allied Trades and
Technicians, London

Committee Members

LAUREN K. SOTH
Editor of the Editorial Pages,
Des Moines Register & Tribune,
Des Moines, Iowa

SIR MICHAEL STEWART
Director, The Ditchley Foundation,
Oxfordshire

RALPH I. STRAUS
Director, R.H. Macy & Company, Inc.,
New York, New York

JAMES A. SUMMER
Vice Chairman, General Mills Inc.,
Minneapolis, Minnesota

HAROLD SWEATT
Honorary Chairman of the Board,
Honeywell Inc., Minneapolis, Minnesota

SIR ROBERT TAYLOR
Deputy Chairman, Standard and Chartered
Banking Group Limited, London

A.A. THORNBROUGH
President, Massey-Ferguson Limited,
Toronto, Ontario

LORD TRANMIRE
Thirsk, Yorkshire

SIR MARK TURNER
Deputy Chairman, Kleinwort, Benson,
Lonsdale Ltd., London

WILLIAM I.M. TURNER, JR.
President and Chief Executive Officer,
Consolidated-Bathurst Ltd.,
Montreal, Quebec

HON. JOHN W. TUTHILL
Director General, The Atlantic Institute,
Paris, France

CONSTANT M. van VLIERDEN
Executive Vice President, Bank of
America, National Trust and Savings
Association, San Francisco, California

R.C. WARREN
Vice President, IBM Corporation, and
President, International Operations
Division, Port Chester, New York

*GLENN E. WATTS
President, Communications Workers
of America, AFL-CIO,
Washington, D.C.

*W.L. WEARLY
Chairman, Ingersoll-Rand
Company, Woodcliff Lake, New Jersey

VISCOUNT WEIR
Chief Executive, The Weir Group Ltd.,
Glasgow, Scotland

HUNTER P. WHARTON
General President, International Union
of Operating Engineers, Washington, D.C.

JOHN R. WHITE
New York, New York

WILLIAM W. WINPISINGER
General Vice President, International
Association of Machinists and
Aerospace Workers, Washington, D.C.

FRANCIS G. WINSPEAR
Edmonton, Alberta

DAVID J. WINTON
Minneapolis, Minnesota

SIR ERNEST WOODROOFE
Formerly Chairman, Unilever Ltd.,
Guilford, Surrey

SIR MICHAEL WRIGHT
Chairman, Atlantic Trade Study and
Director, Guinness Mahon Holdings
Ltd., London

†ARNOLD S. ZANDER
Professor, University of Wisconsin,
Green Bay, Wisconsin

---

*Became a member of the Committee after statement was circulated for signature.

†Deceased.

Sponsoring Organizations

The British-North American Research Association was inaugurated in December 1969. Its primary purpose is to sponsor research on British-North American economic relations in association with the British-North American Committee. Publications of the British-North American Research Association as well as publications of the British-North American Committee are available at the Association's office, 6/14 Dean Farrar Street, London, SW1H ODX (Tel 01-222-6876). The Association is recognized as a charity and is governed by an Executive Committee under the chairmanship of Sir David Barran.

The National Planning Association was founded in 1934 as an independent, private, nonprofit, and nonpolitical organization. It engages in studies and develops recommendations based on nonpartisan research or analysis on major policy issues confronting the United States, both in domestic affairs and in international relations. Its research provides information and methodologies valuable to public and private decision makers.

NPA is governed by a Board of Trustees representing all private sectors of the American economy—business, labor, farm, and the professions. The Steering Committee of the Board, the five Standing Committees (the Agriculture, Business and Labor Committees on National Policy, the National Committee on America's Goals and Resources, and the Committee on International Policy), and special Policy Committees (including the British-North American Committee) originate and approve policy statements and reports. Major research projects undertaken for government and international agencies, and through foundation grants, are carried out with the guidance of research advisory committees providing the best knowledge available. The fulltime staff of the Association totals around 80 professional and administrative personnel.

The Association has a public membership of some 3,000 individuals, corporations, organizations, and groups. NPA activities are financed by contributions from individuals, business firms, trade unions, and farm organizations; by grants for specific research projects from private foundations; and by research contracts with federal, state and local government agencies and international organizations.

NPA publications, including those of the British-North American Committee, can be obtained from the Association's office, 1606 New Hampshire Avenue, N.W., Washington, D.C. 20009 (Tel. 202-265-7685).

Sponsoring Organizations 37

The C.D. Howe Research Institute is a private, nonpolitical, nonprofit organization founded in January 1973 by the merger of the C.D. Howe Memorial Foundation and the Private Planning Association of Canada (PPAC), to undertake research into Canadian economic policy issues, especially in the areas of international policy and major government programs.

HRI continues the activities of the PPAC. These include the work of three established committees, composed of agricultural, business, educational, labor, and professional leaders. The committees are the Canadian Economic Policy Committee, which has been concentrating on Canadian economic issues, especially in the area of trade, since 1961; the Canadian-American Committee, which has dealt with relations between Canada and the United States since 1957 and is jointly sponsored by HRI and the National Planning Association in Washington; and the British-North American Committee, formed in 1969 and sponsored jointly by the National Planning Association, the British-North American Research Association in London, and HRI. Each of the committees meets twice a year to consider important current issues and to sponsor and review studies that contribute to better public understanding of such issues.

In addition to taking over the publications of the three PPAC committees, HRI releases the work of its staff, and occasionally of outside authors, in four other publications: *Observations*, a monthly series; *Policy Review and Outlook*, to be published annually; *Special Studies*, to provide detailed analysis of major policy issues for publication on an occasional basis; and *Commentaries*, to give wide circulation to the views of experts on issues of current Canadian interest.

HRI publications, including those of the British-North American Committee, are available from the Institute's offices, 2064 Sun Life Building, Montreal, Quebec H3B 2X7 (Tel. 514-879-1254).

Publications of the British-North American Committee

BN-17 *Higher Oil Prices: Worldwide Financial Implications,* a Policy Statement by the British-North American Committee and a Research Report by Sperry Lea, October 1975 (£1.50, $3.00)

BN-16 *Completing the GATT: Toward New International Rules to Govern Export Controls,* by C. Fred Bergsten, October 1974 (80p, $2.00)

BN-15 *Foreign Direct Investment in the United States: Opportunities and Impediments,* by Simon Webley, September 1974 (80p. $2.00)

BN-14 *The GATT Negotiations, 1973-75: A Guide to the Issues,* by Sidney Golt, April 1974 (£1, $2.50)

BN-13 *Problems of Economic Development in the Caribbean,* by David Powell, compiled from a study by Irene Hawkins, November 1973 (80p, $2.00)

BN-12 *The European Approach to Worker Management Relationships,* by Innis Macbeath, October 1973 (£1, $2.50)

BN-11 *An International Grain Reserve Policy,* by Timothy Josling, July 1973 (40p, $1.00)

BN-10 *Man and His Environment,* by Harry G. Johnson, May 1973 (40p, $1.00)

BN-9 *Prospective Changes in the World Trade and Monetary Systems: A Comment,* a Statement by the British-North American Committee, October 1972 (30p, $0.75)

BN-8 *Multinational Corporations in Developed Countries: A Review of Recent Research and Policy Thinking,* by Sperry Lea and Simon Webley, March 1973 (80p, $2.00)

BN-7 *Sterling, European Monetary Unification, and the International Monetary System,* by Richard N. Cooper, March 1972 (40p, $1.00) Out of print.

BN-6 *Multinational Corporations and British Labour: A Review of Attitudes and Responses,* by John Gennard, January 1972 (80p, $2.00)

BN-5 *The Strategic and Political Issues Facing America, Britain, and Canada,* by Leonard Beaton, October 1971 (40p, $1.00)

BN-4 *Purposes and Projects: A Policy Statement by the British-North American Committee,* April 1971 (20p, $0.50)

BN-3 *British Entry to the European Community: Implications for British and North American Agriculture,* by John S. Marsh, together with *Agricultural Policies for World Trade Expansion,* a Statement by the British-North American Committee, March 1971 (50p, $1.25)

BN-2 *Transatlantic Relations in the Prospect of an Enlarged European Community,* by Theodore Geiger, November 1970 (60p, $1.50)

BN-1 *An Overall View of International Economic Questions Facing Britain, the United States, and Canada during the 1970's,* by Harry G. Johnson, June 1970 (40p, $1.00)

In Great Britain:
British-North American Research Association
6/14 Dean Farrar Street, London, SW1H ODX.
Telephone: 01-222-6876

In Canada:
C.D. Howe Research Institute
2064 Sun Life Building
Montreal, Quebec H3B 2X7
Telephone: 514-879-1254

In the United States of America:
National Planning Association
1606 New Hampshire Avenue, N.W., Washington, D.C. 20009
Telephone: 202-265-7685

Price: £1.50, $3.00
BN-17

LIBRARY OF DAVIDSON COLLEGE

Books on regular loan may be checked out for **two weeks**. Books must be presented at the Circulation Desk in order to be renewed.

A fine is charged after date due.

Special books are subject to special regulations at the discretion of library staff.

FEB. 15. 1977	FEB 24 1981 FEB 18		
MR. -3. 1977			
NOV. APR. 18 1977 1977 NOV. -6. 1977			
FEB. 25. 1978			
OCT. 19. 1978			
JAN 25 1979 APR. 10. 1979			
JAN 22 1981			
FEB -5 1981			